MECHANIZING
MICROBIOLOGY

Publication Number 1010

AMERICAN LECTURE SERIES®

A Monograph in

The BANNERSTONE DIVISION *of*
AMERICAN LECTURES IN CLINICAL MICROBIOLOGY

Edited by

ALBERT BALOWS, Ph.D.
Director, Bacteriology Division
Center for Disease Control
Atlanta, Georgia

MECHANIZING
MICROBIOLOGY

Edited By

ANTHONY N. SHARPE, Ph.D., F.R.I.C.

Head, Microbiology Automation Section
Bureau of Microbial Hazards
Food Directorate
Health Protection Branch
Health and Welfare
Ottawa, Ontario, Canada

and

DAVID S. CLARK, Ph.D.

Chief, Division of Microbiological Research
Bureau of Microbial Hazards
Food Directorate
Health Protection Branch
Health and Welfare
Ottawa, Ontario, Canada

With a Foreword by

Albert Balows, Ph.D.

Chief, Bacteriology Division
Center for Disease Control
Atlanta, Georgia

CHARLES C THOMAS · PUBLISHER
Springfield · Illinois · U.S.A.

Published and Distributed Throughout the World by
CHARLES C THOMAS • PUBLISHER
Bannerstone House
301-327 East Lawrence Avenue, Springfield, Illinois, U.S.A.

© *1978, by* CHARLES C THOMAS • PUBLISHER
ISBN 0-398-03658-6
Library of Congress Catalog Card Number: 77 2668

With THOMAS BOOKS *careful attention is given to all details of
manufacturing and design. It is the Publisher's desire to present
books that are satisfactory as to their physical qualities and artistic
possibilities and appropriate for their particular use.* THOMAS
BOOKS *will be true to those laws of quality that assure a good
name and good will.*

Printed in the United States of America
N-1

Library of Congress Cataloging in Publication Data

Main entry under title:

Mechanical microbiology.

    (American lecture series, publication ; no. 1010)
    "This book grew out of the International Conference on Mechanized
Microbiology held at Ottawa in September, 1975."
    Bibliography: p.
    Includes index.
    1. Microbiology—Automation—Congresses.   2. Microbiology—Tech-
nique—Congresses. I. Sharpe, Anthony N. II. Clark, David S. III. International Conference on Mechanized Microbiology, Ottawa, 1975.
QR65.M38   576'.028   77-2668

CONTRIBUTORS

N. B. BISCIELLO, JR.
Microbiologist
Food and Drug Administration
Brooklyn, New York

D. W. BLAIR, B.S.
Microbiologist
The Procter and Gamble Company
Cincinnati, Ohio

T. L. BLANEY, Ph.D.
Biochemist
The Procter and Gamble Company
Cincinnati, Ohio

Dr. med. Vet. R. BÖHM
Scientific Assistant
Fachrichtung fur Microbiologie
Institut fur Tiermedizin und
*   Tierhygiene mit Tierklinik*
Universitat Hohenheim
Stuttgart, Federal Republic of
*   West Germany*

W. W. BRINER, Ph.D.
Microbiologist
The Procter and Gamble Company
Cincinnati, Ohio

L. R. BROWN, Ph.D.
Professor of Microbiology and
*   Associate Dean*
College of Arts and Sciences
Mississippi State University
State College, Mississippi

G. E. BUCK, B.S.
Department of Pathology
Utah College of Medicine
Salt Lake City, Utah

P. CADY, Ph.D., M.D.
President
Bactomatic Inc.
Palo Alto, Califorina

J. E. CAMPBELL, Ph.D.
Chief
Microbial Biochemistry Branch
Food and Drug Administration
Cincinnati, Ohio

G. W. CHILDERS, Ph.D.
Research Microbiologist
College of Arts and Sciences
Mississippi State University
State College, Mississippi

I. D. COSTIN, M.D.
Chief
Laboratory for Microbiological
*   Quality Control of*
*   Pharmaceutical Raw Materials*
*   and Culture Media*
Darmstadt, Federal Republic of
*   West Germany*

A. CURTISS, M.M.E.
Manager
Mechanical Engineering
Pfizer Diagnostics Division
Groton, Connecticut

v

J. G. EDWARDS, Ph.D.
Senior Research Microbiologist
Norwich Pharmacal Company
Norwich, New York

E. ENGELBRECHT, M.D.
Chief
Microbiology Dept.
Stichting Samenwerking Delftse
  Ziekenhuizen
Reynier de Graefweg
Delft, Holland

D. FREEDMAN
President
New Brunswick Scientific
  Company
  Inc.
New Brunswick, New Jersey

E. E. GELDREICH, M.S.
Chief
Microbiological Treatment
  Branch
Water Supply Research Division,
  MERL
U.S. Environmental Protection
  Agency
Cincinnati, Ohio

R. GRAPPIN, Ing.Agr.
Ingenieur Principal
Institut National de la Recherche
  Agronomique
Station Experimentale Laitiere
  de Poligny
Jura, France

R. G. HOLCOMB, M.S.
Research Fellow
School of Public Health
University of Minnesota
Minneapolis, Minnesota

W. E. JORDAN, Ph.D.
Microbiologist
The Procter and Gamble Company
Cincinnati, Ohio

D. W. LAMBE, JR., Ph.D.
Director
Regional Diagnostic and
  Developmental Microbiology
  Laboratory of Pathology
New England Deaconist Hospital
Boston, Massachusetts

W. W. LASLIE, M.S.
Supervisor
Regional Diagnostic and
  Developmental Microbiology
  Laboratory of Pathology
New England Deaconist Hospital
Boston, Massachusetts

J. D. MacLOWRY, M.D.
Chief
Microbiology Service
National Institutes of Health
Bethesda, Maryland

J. M. MATSEN, M.D.
Head
Division of Clinical Pathology
Department of Pathology
University of Utah
  College of Medicine
Salt Lake City, Utah

J. McKIE, Ph.D.
Manager
Microbiological Research and
  Development Division
Pfizer Diagnostics
Groton, Connecticut

G. L. MICHAUD, B.S.
Bureau of Microbial Hazards
Food Directorate
Health Protection Branch
Health and Welfare
Ottawa, Ontario, Canada

T. E. MUNSON, B.S.
Microbiologist
Food and Drug Administration
Brooklyn, New York

J. J. PARRAN, M.S.
Microbiologist
The Procter and Gamble Company
Cincinnati, Ohio

I. J. PFLUG, Ph.D.
Professor, Food Science and
Nutrition
Department of Food Science and
Nutrition
University of Minnesota
Minneapolis, Minnesota

R. B. READ, JR., Ph.D.
Acting Director
Division of Microbiology
Bureau of Foods
Food and Drug Administration
Washington D. C.

D. J. REASONER, Ph.D.
Research Microbiologist
Microbiological Treatment
Branch
Water Supply Research Division,
MERL
U.S. Environmental Protection
Agency
Cincinnati, Ohio

E. A. ROBERTSON, M.D.
Assistant Chief
Laboratory Computer Service
National Institutes of Health
Bethesda, Maryland

J. P. SCHRADE, B.S.
Supervising Microbiologist
Food and Drug Administration
Brooklyn, New York

J. SEO, Ph.D.
Senior Scientist
Department of Research and
Development
Pfizer Diagnostics
Groton, Connecticut

A. N. SHARPE, Ph.D., F.R.I.C.
Head
Microbiology Automation Section
Bureau of Microbial Hazards
Food Directorate
Health Protection Branch
Health and Welfare
Ottawa, Ontario, Canada

B. H. SIELAFF, Ph.D.
Project Director
Pfizer Diagnostic Division
Groton, Connecticut

R. E. TROTMAN, M.Sc., Ph.D
C.Eng., M.I.E.E., F.Inst.P.
Director
Bio-Engineering Department
St. Mary's Hospital Medical
School
London, England

J. S. WINBUSH, M.S.
Acting Director
Division of Mathematics
Food and Drug Administration
Washington D.C.

J. A. WUNDER, B.A.
Microbiologist
The Procter and Gamble Company
Cincinnati, Ohio

FOREWORD

The GENESIS of this series, *The American Lecture Series in Clinical Microbiology,* stems from the concerted efforts of the Editor and the Publisher to provide a forum from which well-qualified and distinguished authors may present, either as a book or monograph, their views on any aspect of clinical microbiology. Our definition of clinical microbiology is conceived to encompass the broadest aspects of medical microbiology not only as it is applied to the clinical laboratory but equally to the research laboratory and to theoretical considerations. In the clinical microbiology laboratory we are concerned with differences in morphology, biochemical behavior, and antigenic patterns as a means of microbial identification. In the research laboratory or when we employ microorganisms as a model in theoretical biology, our interest is often focused not so much on the above differences but rather on the similarities between microorganisms. However, it must be appreciated that even though there are many similarities between cells, there are important differences between major types of cells which set very definite limits on the cellular behavior. Unless this is understood it is impossible to discern common denominators.

We are also concerned with the relationships between microorganism and disease—any microorganisms and any disease. Implicit in these relations is the role of the host which forms the third arm of the triangle: microorganisms, disease, and host. In this series we plan to explore each of these; singly where possible for factual information and in combination for an understanding of the myriad of interrelationships that exist. This necessitates the application of basic principles of biology and may, at times, require the emergence of new theoretical concepts which will create new principles or modify existing ones. Above all, our aim is to present well-documented books which will be informative, instructive, and useful, creating a sense of satisfaction to both the reader and the author.

Closely intertwined with the above *raison d'être* is our desire to produce a series which will be read not only for the pleasure of knowledge but which will also enhance the reader's professional skill and extend his technical ability. The *American Lecture Series in Clinical Microbiology* is dedicated to biologists—be they physicians, scientists, or teachers—in the hope that this series will foster better appreciation of mutual problems and help close the gap between theoretical and applied microbiology.

Attention has been sharply focused over the past decade on the development of automated instruments which lend themselves to one or more time, labor, or material saving aspects of laboratory effort. Simultaneously, automated or semiautomated analytical devices have also been developed in the laboratories of innovative researchers. As a result of these and related activities, a number of instruments, devices, and methodology improvements have occurred in several areas within the broad scope of microbiology.

There has been at least one European international conference designed specifically to provide a forum for the presentation and discussion of rapid methods and automation as applied to microbiology. However, no such meetings had been organized for the Western Hemisphere until Dr. A. N. Sharpe and Dr. D. S. Clark of the Health and Welfare Branch of the Canadian Ministry of Health and Welfare put such a conference together. The conference was held in Ottawa in the autumn of 1975. It was international in scope and broad in the coverage of mechanization in microbiology. The meeting was well received and the organizers of the conference were urged to publish the proceedings. Thus, Drs. Sharpe and Clark assumed the role of editors and made possible this addition of the *American Lecture Series in Microbiology*. It is a most welcome addition and one which should serve not only as a source of baseline information and data on certain aspects of mechanization but also to serve as an inspiration to those readers who will be motivated to improve on these instruments described in this book so that microbiology will continue to keep time with the new drummer—automation.

ALBERT BALOWS, PH.D.
Editor

PREFACE

THIS book grew out of the International Conference on Mechanized Microbiology held at Ottawa in September, 1975. We hoped that by inviting conference contributors to rewrite their manuscripts in the light of the general outcomes of the conference, instead of simply publishing a verbatim conference report, a more coherent and valuable work would be obtained. By so doing, we also enabled many of the contributors to widen the scope of their papers, particularly by the inclusion of more review material.

Two other chapters, not originally presented as papers at the conference, have been added. Chapter 1, by R. E. Trotman, describes in very readable terms some of the frustrations, problems, and prejudices surrounding the whole business of mechanizing microbiology. Chapter 3, by A. N. Sharpe, attempts to show that the detection of microorganisms, contrary to the opinions of many microbiologists, is subject to the same laws as the detection of anything else. Other sciences have benefited from the application of theories of communication, and the further exploration of microbiology by such means may well uncover processes inherently more suited to mechanization than those we use at present.

The would-be inventor should find several areas described within the book from which useful developments might be made. The field of microbiology, particularly food microbiology, badly needs new ideas, new enthusiasms, and strong research and development investment. Chapter 5, by R. B. Read, paints a slightly gloomy picture. However, the scientist-inventor should find in Chapter 4, by D. Freedman, a friendly encouragement, some useful guides and, for those who may have seen their brainchildren dashed against the rocks, some appreciation that inventions are not lightly tossed into the sea of commerce.

A.N.S.
D.S.C.

CONTENTS

MECHANIZING
MICROBIOLOGY

Chapter 1

MECHANIZING MICROBIOLOGY: THE ADVANTAGES AND FUTURE PROSPECTS

R. E. TROTMAN

IT IS usually claimed that the primary advantages of mechanization in any discipline are that it increases productivity and that it relieves human beings of having to perform soul-destroying, menial, and repetitive tasks (sometimes necessarily performed in unhealthy environments) so that they may be released to perform mainly those tasks requiring the special training and skills only the technologist possesses. Those are undoubtedly some of the advantages of mechanizing microbiology (except perhaps in the developing countries in which there is a profusion of unskilled labor). However, despite the many attempts to devise and introduce automatic methods into routine use, even simple aids such as semiautomatic dilutors, dispensers, and turbidometric devices, as well as more sophisticated techniques based on the computer or on impedance or on differential light-scattering measurements, are still used sparingly in medical and other branches of microbiology. Those techniques will be discussed later in this volume, and in Hedén & Illéni (1975) and in Trotman (1977), but the question of why this is so will be examined here.

One reason sometimes advanced is that the technology is not available. It is of course true that it is not yet possible to, for instance, identify the organisms in a mixed culture without first isolating them, although claims that one can do this with some techniques, such as impedance measurements, differential light scattering, and gas-liquid chromatography, have been made. However there is a great deal of technology available. Very much is technically feasible.

Of course, it does not follow that, because development of a

3

specific technique and/or apparatus is technically feasible, it is a useful objective; many designers, both amateur and professional, have fallen into the trap of devising very ingenious but rarely required methods and devices. Some designers have attempted, and failed, to overcome the many technical problems involved, especially in mechanically handling infected material. Furthermore, badly designed, unreliable, costly to run, and misapplied equipment, of which, unfortunately, one knows so many examples, will put an apparatus into disrepute, even though in principle it is very valuable, thereby raising doubts about its value and that of similar equipment. This encourages the belief that it is not possible to produce useful automatic methods in microbiology. Additionally, designers have produced apparatus capable of such outputs that only a few machines would be required to perform all the work in a country of the size of the United Kingdom. These practices waste resources and can permanently discourage a microbiologist from introducing mechanization. One should not rush in and develop one's brainchild without having first firmly established that it really is potentially a tool that is *needed* and likely to be economic when *all* overheads, not just the cost of the capital equipment and consumables, are taken into account.

We are far from having produced all the feasible and practical methods and devices for use in routine laboratories and are far from overcoming the difficulties of introducing good equipment into routine use. The latter is a major problem arising partially because there are those who doubt that mechanizing microbiological laboratories, even if technically feasible, serves a useful purpose. But there is no doubt that there are many advantages, in addition to the primary ones outlined above, in using automatic methods, provided equipment is designed to perform a specific microbiological function. Modifying equipment designed for a different application can be valuable but is often carried out badly and inappropriately.

The variations in the results obtained by two or more workers, ostensibly carrying out identical test procedures, even in the same laboratory, are well known (Gavan, 1974). Much of this variation can be eliminated by the use of a well-designed and constructed machine, provided it is functioning properly, because it is more consistent than human beings. Furthermore, the contribution of the observer error to the total variation is significantly reduced. In addition, machines can be more sensitive and more accurate than human beings. We must stop such practices as holding a culture up to the light and saying "Oh yes, that's about 10^6 organisms per ml." In this day and age a more scientific approach should be the norm.

An additional advantage of automatic methods is that the results obtained from machines are readily sent directly to data processing equipment. Furthermore, mechanized equipment can be programmed, and it can and should be designed in such a way as to indicate when parameters are outside predetermined limits, when it is out of sequence, and when it has failed. However, one must not overlook the fact that even well-designed and well-constructed equipment needs routine calibration, maintenance, and servicing, and that its proper use requires an operator with appropriate aptitudes.

There is at present very little quality control in microbiology, although the situation is beginning to improve. The use of automatic apparatus facilitates much wider use of such control, although control of media and reagents is required even in manual methods (Russell et al., 1969).

The above seem to be very important reasons for introducing automatic methods into microbiology, but are they in themselves adequate to justify the investments needed to design, develop, and produce such methods? Bearing in mind the harsh world in which we live, can automatic methods be more than bonuses, the main or sole purpose of introducing them being to increase productivity?

One wonders whether the cost of developing automated equipment is a justifiable deterrent, because the bases for economic comparisons are difficult to determine. For example, it would be very difficult to establish the cost of, say, a false diagnosis due to an error in performing a manual test (although some industrialists no doubt have a shrewd idea). Also, is not an improvement in the quality of service to clinicians or food technologists, ensuing from the introduction of well-designed automatic methods, sufficient justification in itself? Perhaps the answer depends on the particular field of endeavour.

Productivity is more easily related to cost-effectiveness, although this author knows of no studies on the cost-effectiveness of introducing automatic techniques into any type of microbiological laboratory. No doubt, as more techniques are introduced, such studies will be carried out. Nevertheless, in the author's experience, the sensible application of automatic techniques improves the productivity of the laboratory and the quality of the microbiological service. However, probably comparatively few microbiologists are so convinced.

If they are not, those who have attempted to devise automated equipment must shoulder some of the responsibility for not being more convincing. But, a more flexible and less prejudiced approach is required of microbiologists in general. As Stafford (1968) suggested, the "production-line" attitude of mind is alien to the scientist but may well be necessary in the future. Clinical chemists are now used to such methods, and it is doubtful that microbiologists would find thinking along similar lines too traumatic.

It may well be that one could remove the doubts that microbiologists may have about the value of automated microbial methods if sufficient good quality machines (prototypes of production models) were available to enable a substantial number of microbiologists to have one on trial long enough to assess its advantages. They would probably find that many of their fears are unfounded. Experience suggests that it is not difficult to demonstrate to the microbiologist the advantages of introducing automatic methods once he has had the opportunity to personally evaluate a good device. Indeed, he usually becomes most enthusiastic. The present lack of enthusiasm is a major obstacle to progress and can only be overcome if the following unfortunate, but seriously inhibiting, vicious circle can be broken out of.

At present a commercial firm looking at the field sees very little encouragement to invest in it. As far as can be ascertained, the potential market for any new hardware appears to be very limited. The firm is not, therefore, prepared to invest money, even to the extent of producing a few prototypes of potentially useful devices, especially if it is unlikely to obtain patent protection. So many valuable ideas and devices are unfortunately not patentable, because existing technology is applied, as has been explained (Trotman and Byrne, 1975).

However, the true market is probably grossly underestimated because few microbiologists can truthfully and enthusiastically say in advance that they would purchase a proposed apparatus, primarily because they have had no opportunity to evaluate the apparatus and are not, therefore, aware of its potentialities. But microbiologists can only have an opportunity to assess equipment if at least a few prototypes are available. How then can we break out of this vicious circle and produce these prototypes?

This is a particularly difficult area, one that has troubled many people for many years. The difficulties in translating ideas and laboratory prototypes into commercial prototypes and products are well known and not confined to medical laboratory technology or, for that matter, to bioengineering. The United Kingdom National Research and Development Corporation (NRDC) was established to deal with this problem, by attempting to bring scientists-inventors working in noncommercial organizations into contact with appropriate commercial manufacturers, for the purpose of developing, manufacturing, and marketing their ideas.

While some well-known and successful projects have been pioneered by the NRDC in conjunction with inventors, many useful ideas and laboratory prototypes never see the light of day, because very few people other than the original inventors are able to use them. There is a feeling that the NRDC fails to help in many cases, and in 1971 Walker suggested that the corporation, which usually acts as a middleman, may actually be a deterrent to the exploitation of inventions. However, one cannot blame the NRDC for all the difficulties in developing and producing automated microbiological methods, because only a handful of ideas have been taken to it in the last few years. Furthermore, all inventors think their ideas are world shattering, and the NRDC has the unenviable task of disillusioning many; in the process it is blamed for not doing the job properly.

Direct liaison between the inventor and an appropriate commercial organization is a well-tried method and one advocated by Walker (1971). However, although this approach has undoubtedly been profitable in some cases, the world is littered with relics of unsuccessful attempts at such liaisons. Similarly, many firms have been specially created to develop and market a new device, but, alas, despite some notable successes, the world is also littered with relics of these attempts.

It is clear, therefore, that if more good designers are going to be encouraged to devote their energy to developing automated microbiological methods, and if there is going to be more progress made in introducing such methods into routine use, a new approach is needed to the problem of translating new ideas and laboratory prototypes into commercial products.

It seems that there is a need for a biological/engineering unit attached to an institution in which research and development in microbiological methods is taking place, the primary function of which would be to assess the various developments in progress in research laboratories and to produce a small number of prototypes of potentially valuable machines. These prototypes would be used primarily to assess their reliability, effectiveness, and usefulness in the field. It is very important that this unit be attached to an applied research institution rather than to an industrial organization, because direct access to, and acquaintance with, the field is essential. But the unit must have on its staff engineers widely experienced in industrial manufacturing processes, who are familiar with all the constraints under which industrial organizations have to work, since the end product of a project will hopefully be manufactured and marketed. But an arrangement must be made to ensure that they do not lose touch with those constraints. Industrial contacts are just as important as microbiological contacts in this concept. Secondment may be the answer.

It needs but one organization. After all, one organization changed the face of clinical chemistry by the development of the Auto-analyser.® In 1957, when Skeggs first published his work that led to the development of the Auto-analyser, many clinical chemists no doubt said much the same as some microbiologists are saying today—"A machine cannot do it: it takes far too much skill and many years of training;" or "You cannot be sure that adequate safeguards against error and false sample identification can be taken in a machine;" or "What about contamination?" But it is now known that such fears are unfounded. Machines can perform many functions satisfactorily, and one can build more safeguards into a machine than are built into human beings.

Sufficient effort has not yet been devoted to overcoming the technical and other problems involved in developing and introducing mechanization into routine use. Somebody, somewhere, someday, must surely take the plunge and reap the reward for far-sightedness.

REFERENCES

Gavan, T.L.: 1974. A summary of the bacteriology portion of the 1972 basic, comprehensive and special College of American Pathologist (CAP) Quality Evaluation Program. *Am Clin Pathol Suppl, 61*:971.

Hedén, C.G. and T. Illéni (eds.): 1975. *Automation in Microbiology and Immunology and New Approaches to the Identification of Microorganisms.* Wiley, New York.

Russell, R.L., R.S. Yoshimori, T.F. Rhodes, J. W. Reynolds, and E.R. Jennings: 1969. A quality control program for clinical microbiology. *Am Clin Pathol, 52*:489.

Skeggs, L.J.: 1957. Automatic method for colorimetric analysis. *Am Clin Pathol, 28*:311.

Stafford, J.L.: 1968. Tomorrow's laboratory. *Proc R Soc Med, 61*:1047.

Trotman, R.E. and K.C. Byrne: 1975. The automatic preparation of bacterial culture plates. *J Appl Bacteriol, 38*:61.

Trotman, R.E.: 1977. *Technological Aids to Bacteriology.* Acad Pr, London. In press.

Walker, P.M.B.: 1971. Rewards for inventors. *Nature, 231*:357.

Chapter 2

A SURVEY OF POSSIBILITIES FOR MECHANIZATION OR AUTOMATION OF MICROBIOLOGICAL PROCEDURES

R. BÖHM

Abstract

A short survey on methods of automation and mechanization in microbiology is given. An evaluation of the methods results in the following grouping: quantitative methods with direct and indirect determination of the bacterial count; qualitative methods on the basis of biochemical parameters, physiochemical procedures, and serological reactions. It was found that diagnostic methods applicable to single bacterial cells should be developed.

Introduction

T HE following paper surveys the possibilities of automation in microbiology. All the different approaches are presented without any evaluation of their applicability. All these methods, except some quantitative ones for special purposes, are still in the experimental stage. It should be pointed out that a lot of different developments have been initiated, and encouragement should be given for their improvement and perfection. Some guidelines for further work should also be provided in order to arrive at practicable routine methods.

The Purposes of Mechanization and Automation in Microbiology

Automation and mechanization in microbiology are not introduced primarily to save personnel. At least two further points are of interest. The first is to liberate the technical personnel from simple repetitive work, especially in some quantitative methods. The second is especially important in the field of bacteriological diagnosis and taxonomy. Here, an ever increasing number of diagnostic and taxonomic parameters makes electronic data processing necessary because the analysis would make high demands on the individual. This and the modern understanding of biology have led to the application of probabilistic methods in the field of microbiology, which requires special methods of procuring data.

Theoretical Aspects of the Rapid Identification of Bacetria

Conventional methods of identifying bacteria are based either on the metabolism of the microorganism or on the serological reaction of its antigens. The judgment as to whether or not a substrate is metabolized is influenced by the following factors:

the nature of the culture medium in use

the preparation of the medium

the type of and preparation of any necessary indicator system

duration of incubation

the amount and method of inoculation

the experience and skill of the personnel involved

Because all of these variables enter into the identification it is not surprising that Sneath (1973) found the percentage of error within one laboratory to be at most 4 percent, while between different laboratories it was 8 to 15 percent. The use of serological procedures has been successful for certain groups of bacteria and forms a basis for identification, whereas for other groups, because of cross-reactions, it is only partly or not at all applicable. These procedures are, in general, expensive and are also not completely free from subjective opinion. Because of the disadvantages of the conventional methods, there arises a demand for a system of rapid identification which can be automated. The method must be able to be internationally standardized and should, because of the danger of changing the basis of evaluation through mutation, avoid considerations based on metabolites. In addition, the measured parameters should be amenable to conversion into a form suitable for data processing.

A Survey of Current Methods and Procedures

An evaluation of the apparatus and procedures developed to date results in the following grouping:

A. *Quantitative procedures*
   1. Methods for the direct determination of the bacterial count.
   2. Methods for the indirect determination of the bacterial growth.

B. *Qualitative procedures*
   1. Rapid identification methods which depend primarily on the biochemical metabolism of the bacteria.
   2. Rapid identification methods which depend predominantly on physiochemical methods.
   3. Rapid identification methods which are based predominantly on serological reactions.

Quantitative Methods

Direct Methods

Initially, the quantitative measurement of the bacterial growth naturally lent itself to accelerated techniques and/or automation. The need to prepare serial dilutions has led to two different approaches. One is miniaturization with the help of the microtiter system, as described by Fung in 1973, and the other the use of the automatic pipetting machine as, for example, the one for solid culture media as described by Sharpe et al. (1972) or the Bioreactor® for liquid media as described by Engelbrecht in 1974. These systems can also be put to use in other areas of microbiology, as in the case of the susceptibility tests described by Isenberg in 1973, in a procedure based also on miniaturization, and by Zijlstra and Dankert (1974) with the Bioreactor. For the direct method of bacterial count in liquids, the Coulter Counter® has already been proved by Tolle and Zeidler (1968) to be useful in certain instances. For the automatic bacterial plate count there are many electronic counting devices available, such as the Artec Colony Counter®, or the spiral platecounter based on the laser techniques as described by Read et al. (1974).

Indirect Methods

For the indirect method of bacterial count, procedures have been investigated recently which employ the continuous-flow-system (Auto-analyzer) so valuable in clinical chemistry. In 1974, Heeschen reported the possibility of using pyruvate production as an indication of the bacterial growth in milk, while Hamann, in 1974, also referred to the Auto-analyzer as a means of carrying out the suspectibility test making use of the breakdown of glucose as its parameter.

Qualitative Methods

Methods Based on Biochemical Reactions

The qualitative rapid identification procedures which depend predominantly on biochemical metabolism can be differentiated into the two following groups: those miniaturized tests which are read directly and evaluated in the conventional manner, and those methods which automatically register and, in some instances, evaluate certain parameters of the biochemical capacities of the bacteria. Within the miniaturized systems one can distinguish between the commercial and the self-constructed ones based on the microtiter system. The commercial systems, which tie the investigator down to a certain, fixed scheme, come on the market either in the form of small pre-prepared plastic reagent-containing reservoirs as in the API-system or as test-tablets or paper strips which have been impregnated with the reagents. The miniaturizations recommended by Fung (1972), based on the microtiter system have, according to evidence from Hahn (1974), proved to be successful. In addition to being less expensive, they adapt better to special investigative problems. The second group contains a number of varying methods, the main characteristic being the presence of a common, easily measured final step which produces a more or less diagnostically valuable signal. The pattern of differentiation is obtained either through a number of samples to which, if necessary, different supplements have been added, or through the analytical investigation of one complex sample. Wagner et al. (1973) present the simplest method of this kind, whereby bacterial growth is indicated by the formation of $^{14}CO_2$ from ^{14}C-containing substrates. The system could be controlled by the addition of antibody and used as a quantitative measure of immunological reaction. Cady (1973) presents another principle whereby bacterial growth can be measured within one hour, in which impedance changes in the culture medium are measured. The addition of antibodies and growth-inhibiting substances could make identification through a "negative profile" possible. In the procedure described by Bascomb and Grantham (1973) the results are obtained within two hours. Here the signal is produced by ammonia and CO_2 production through the enzymatic conversion of different substrates. Fifteen substrates serve as the differentiation core. According to these workers (1974) this method can be completely automated by employing the Auto-analyser.

Physiochemical and Physical Methods

Gas chromatography belongs to the category of rapid tests which predominantly depend on physiochemical procedures. It includes the possibility of directly analyzing not only the substrate but also the bacteria, as shown by Mitruka and Alexander in 1972, Mitruka (1973), and many others. In addition to metabolites from the culture medium, organisms infused with serum are also being investigated for specific changes. The procedure should also be suitable for the identification of mixed cultures of up to three to four different bacteria. The preparation of the sample for gas chromatography proceeds automatically, too, and the signals received from the detector are electronically processed, analyzed, and stored.

Gas chromatographic techniques can also be used for the analysis of the cellular components of bacteria. This method has et al. (1972), Brooks et al. (1972), Mitruka and Alexander (1972), been in use for some time by many teams, for example, Bøvre and Bennett and Asselineau (1970), in order to clarify taxonomical questions and to explain cellular structure.

Another valuable method for the structural investigation of bacteria could be infrared spectrophotometry. This has been used by Ivanov and Uvarova (1967), Scopes (1962), and Parnas et al. (1966) in their experiments. A means of group identification is offered by this method.

Recently, pyrolysis mass spectrometry has been recommended for bacterial differentiation. Meuzelaar and Kistemaker (1973) reported a high reproducibility of data using this technique to analyze bacterial colonies on solid medium.

As a purely physical method, light scattering has also been considered as a means of identification. In 1968, Wyatt described how he could distinguish *Bacillus anthracis, Bacillus subtillis, Escherichia coli* and *Staphylococcus aureus* from each other. Furthermore, in 1972, Wyatt and Phillips made use of this procedure in order to explain aspects of bacterial structure.

A completely different principle underlies the attempts for the rapid identification of bacteria by means of fluorescent-spectrophotometry. According to Ginell and Feuchtbaum (1972), *E. coli* and *Sarcina lutea* could be distinguished by means of their absorption and fluorescence spectra. Experiments have indicated that these methods could also be expanded to include the area of microspectrophotometry in order to divide bacteria into several diagnostical groups (Böhm, unpublished results).

Serological Methods

The serological procedures often provide rapid test methods in themselves, as for example the fluorescent-serological method or slide agglutination. Great efforts have been made to automate identification with fluorescent antibodies. Fully automated systems of this kind have been developed in the United States in which the samples are automatically put on a conveyor belt or on plastic slides supplied with conjugate, washed, and evaluated. Such instruments are described by Kaufman et al. (1971) and by Mansberg and Kusnetz (1966).

Conclusion

Many new ways for diagnostic methods have been suggested, for which subsequent work will lead to practicable methods. Some critical remarks must be added, however. Most of these diagnostic procedures do not represent very rapid methods, since they only shorten the diagnostical step. The most time-consuming work of isolating a pure culture has still to be done. Only a few really rapid methods exist, such as those which use labelled antibodies, because most of these do not require any purification of the culture. Efforts should be made to develop single-cell methods. Perhaps the transformation of histochemical methods common in cytology will lead to some success. A further warning must be made against the application of methods already developed to other fields of microbiology without critical examination since, as has been shown by Garttner et al. (1975), the different nature of samples to be examined often represents the source of a high percentage of error.

Finally, it can be stated that the development of a universal diagnostic procedure is not to be expected within the near future.

REFERENCES

Bascomb, S. and C.A. Grantham: "Specific Enzyme Profile," an Automated Method for Bacterial Classification. *1st Int Congr Bact Jerusalem,* Abstracts II, 132.

Bascomb, S. and C.A. Grantham: 1974. Rapid Identification of Bacteria by Continuous Flow Analyses. *IAMS—IX Int Symp Kiel,* Abstracts 21.

Bennet, P. and J. Asselineau: 1970. Influence de l'age sur la teneur en acides gras a chaine ramidiée du bacille tuberculeux. *Ann Inst Pasteur, 118:*324.

Bøvre, K., R. Hytta, E. Jantzen, and L.O. Fronholm: 1972. Gas chromatography of bacterial whole cell methanolysates. *Acta Path Microbiol, Scand (B), 80:*683.

Brooks, J.B., R.E. Weaver, H.W. Tatum, and S.A. Billingsley 1972. Differentiation between *Pseudomonas testosteroni* and *P. acidovorans* by gaschromatography. *Can J Microbiol, 180:*1477.

Cady, P.: 1973. Impedance changes in media as a rapid method of identifying microorganisms. *1st Int Congr Bact Jerusalem,* Abstracts II, 130.

Engelbrecht, E.: 1974. The Bioreactor and Biodilutor as tools for the automation of tests to be performed on serial dilutions of samples. *IAMS— IX Int Symp Kiel,* Abstracts 35.

Fung, D.Y.C.: 1972. Biochemical test-miniaturized. In *Handbook of Microtiter Pocedures,* pp. 40, Dynatech Corp, Cambridge, Mass.

Garttner, E., W. Muller, and F. Farmanara: 1975. Die Zahlung von Luftkeimokolonien mit einem elektronischen Zahlgerat (Colony Counter).

Zentralbl Veterinaermed B, 22:326.

Ginell, R. and R.J. Feuchtbaum: 1972. Fluorescent spectrophotometry in the identification of bacteria. *J Appl Bacteriol, 35*:29.

Hahn, G.: 1974. Rapid identification of streptococci. *IAMS—IX Int Symp Kiel*, Abstracts 33.

Hamann, J.: 1974. The analytics of antibiotic inhibitors in milk. *IAMS—IX Int Symp Kiel*, Abstracts 38.

Heeschen, E.: 1974. Determination of metabolites of bacterial activity in milk. *IAMS—IX Int Symp Kiel*, Abstracts 15.

Isenberg, H.D.: 1973. Antibiotic Methods. *1st International Congress for Bacteriology, Jerusalem*, Abstracts I, 174.

Ivanov, K.K. and R.N. Uvarova: 1967. Investigation of the antigens of paratyphoid B. bacteria by infrared spectrophotometry. *Biofisika, 6*:973.

Kaufman, G.J., J.F. Nester, and D.E. Wasserman: 1971. An experimental study of lasers as exitation sources for automated fluorescent antibody instrumentation. *J Histochem Cytochem, 19*:469.

Mansberg, H.P. and L. Kusnetz: 1966. Quantitative fluorescence microscopy fluorescent antibody automatic scanning techniques. *J Histochem Cytochem, 14*:260.

Meuzelaar, H.L.C and P.G. Kistemaker: 1973. A technique for fast and reproducible fingerprinting of bacteria by pyrolysis-mass-spectrometry. *Anal Chem, 45*:587.

Mitruka, B.M. and H. Alexander: 1972. Halogenated compounds for the sensitive detection of clostridia by gas chromatography. *Can J Microbiol, 18*:1519.

Mitruka, B.M.: 1973. Rapid Automatic Identification of Bacteria. *1st Int Congr Bact, Jerusalem*, Abstract I, 173.

Parnas, J., S. Poplawski, and M. Cybulska: 1966. Investigations on serotypes of Leptospirae by infrared spectrophotometry. *Z Immunitaetsforsch, 132*:218.

Read, R.B., J.E. Gilchrist, and J.E. Campbell: 1976. Spiral method for plating and counting bacteria. *IAMS—IX Int Symp Kiel*, Abstracts 3.

Scopes, A.W.: 1962. The infrared spectra of some acetic acid bacteria. *J Gen Microbiol, 28*:69.

Sharpe, A.N., D.R. Biggs and R.J. Oliver: 1972. Machine for automatic bacteriological pour plate preparation. *Appl Microbiol, 24*:70.

Sneath, P.H.A.: 19773. Test reproducibility in relation to identification. *1st Int Congr Bact, Jerusalem*, Abstracts II, 87.

Tolle, A. and H. Zeidler: 1968. Die elektronische Mikrokoloniezahlung—ein Verfahren zur Beurteilung der bakteriologisch-hygienischen Qualitat der Rohmilch. *Milchwissenschaft, 23*:65.

Wagner, H.N., M. Chen, and S.M. Larson: 1973. Early detection of bacterial growth. *1st Int Congr Bact, Jerusalem*, Abstracts I, 172.

Wyatt, P.J.: 1968. Differential light-scattering, a physical method for identifying living bacterial cells. *Appl Optics, 7*:1879.

Wyatt, P.J. and D.T. Phillips: 1972. Structure of single bacteria from light scattering. *J Theor Biol, 37*:493.

Zijlstra, J.B. and J. Dankert: 1974. Determination of antimicrobial susceptibility by dilution with a semi-automated microtechnique (Bioreactor). *IAMS—IX Int Symp Kiel,* Abstracts 40.

Chapter 3

SOME THEORETICAL ASPECTS OF MICROBIOLOGICAL ANALYSIS PERTINENT TO MECHANIZATION*

A. N. SHARPE

Abstract

This article explores and describes aspects of microbiological analysis, using some of the simpler concepts of information and communication theories. The unique sensitivity of conventional microbiological analyses inherent in the property of multiplication and the minimum quantities of information required in various microbiological methods are described. On the basis of these considerations, a possible reason is given for doubting the practical performance of electronic colony counting methods. The possible format of a satisfactory machine for quantitative microbiological analysis is derived.

Introduction

THE responsiveness of living organisms to environmental variations is particularly evident to the microbiologist. With rapid rates of reproduction and a tendency to produce spontaneous genetic changes, microorganisms have a great ability to resist prolonged attempts at precise definition and control. Of necessity, the science of experimental microbiology has developed, over the years, methods that demand a uniquely human interpretative ability and judgment derived from experience.

For the automation scientist, and particularly the commercial developer of automated microbiological instruments, the need for human judgment in experimental methods is quite a hindrance,

---

*I am grateful to Mr. M. Diotte and Dr. G. Jarvis for much helpful advice and argument.

since human abilities and the detection of the observational associations on which recognitive cues have been built up are often far from being imitated successfully by machine. For example, a dense, shiny black colony, surrounded by a white margin and a zone of cleared Baird-Parker's agar is associated, by the food microbiologist, with the likely presence of enterotoxin-forming *Staphylococcus aureus* contamination. Nothing, however, about the physical appearance of the colony is directly related to the existence of enterotoxin in the food specimen. Methods involving a messy sequence of events by means of which the diagnosis is made—weighing, blending, diluting, preparation and inoculation of agar plates, incubation, inspection, followed perhaps by subculturing and carrying out "confirmatory tests"—have become firmly entrenched and are virtually unassailable by automation scientists.

The remarkable effectiveness of microbiological growth methods over conventional physiochemical methods of attacking analytical problems has strengthened the entrenchment against automation. Microbiology stands out like a sore thumb against those sciences where costly instruments carry out analytical procedures accurately and reliably. In order to analyse 10^{-12}g quantities of a material, a mass-spectroscopist will use a forbiddingly expensive mass-spectrometer, an X-ray spectroscopist, an equally expensive microanalyser, a radiochemist, a scintillation counter, and so forth through all of the physical sciences. But a microbiologist will ask for a bottle or two of medium and a few Petri dishes, knowing that he can make the microorganisms themselves serve as a detector and amplifier of a sensitivity unequalled in the physical world. The basic ease with which humans manipulate the microbial amplifier and detect its output, e.g. in the form of colonies, is one of the major obstacles to the development and introduction of automated microbiological instruments. The difficulties of automating or mechanizing microbiological analyses are so severe that useful developments will now probably only be made if microbiologists can be persuaded to put aside conventional ways of approaching the subject in favor of new methods more suited to the abilities of achievable automation.

Analytical techniques in microbiological methodology have been developed by human intuitive means, rather than by theoretical exploration—a consequence, no doubt, of the ability of the human mind to perceive routes and objectives in the absence of hard data. The outcome has been the development of many closely related analytical themes, the survival of any method depending on its fitness. Some of the results of this evolutionary approach are now better described in terms derived from other sciences or with mathematics. One can reasonably assume, for example, that the accuracy of a single measurement depends inversely on the quantity of information sought; small quantities of information can be obtained much more reliably than can larger quantities. Thus, there is generally an answer to the question, "Does the sample contain *Salmonella derby?*" with greater confidence than the question, "How many cells of *Salmonella derby* does the sample contain?" To answer the second question requires much more information* than the first.

It is sometimes valuable to explore one science using terms taken from another. It may then be found that familiar concepts take on a new significance, if only through being phrased differently. Microorganisms are subject to the same physical and chemical laws observable in the more definable sciences; their apparently capricious behavior from time to time results from the interaction of variables over which we do not yet have sufficient control. Perhaps if we are able to show that many aspects of microbiology are quite similar to those of the more definable sciences, the subject will be less disconcerting to many scientists whose contributions, particularly in the field of mechanization or automation, could be so valuable to microbiologists. The remainder of this chapter describes some of the effects of microbiology in terms derived mainly from the sciences of electronics and communications, in hopes of stimulating some fresh looks at microbial methodology. At the least, one may obtain a new viewpoint of microbial detection. At the best, one may be able to define detection procedures more suited to particular applications,

---

*The term *quantity* is used here in the context of information theory and should not be confused with *value*.

such as mechanization. Purists of a pillaged science will argue that their terms have precise and restricted meanings, and that their use out of context is to invite disaster. That may be so; nevertheless, many of the basic ideas carry through, and even though the use of terms taken from communication theories to describe effects in microbiology may give them a degree of imprecision, or blurring, the general meanings should be quite obvious.

Information and Entropy

It is generally accepted that the entropy (or, more strictly, negative entropy) of a system is a measure of its information content. The first mathematical treatment of information in this way was described by Shannon (Shannon and Weaver, 1949), and consolidated by Brillouin (1956). Useful introductions in information theory can be found in Hancock (1961), Pierce (1961), Rosie (1966), Goodyear (1971), and Young (1971). Every physical, chemical, or biochemical process involving a transfer of energy also involves a change in entropy and a flow of information. Consider an initial situation in which P different but equally probable things might happen, passing to one particular and certain outcome as a result of information being received. For example, a question which has P different and equally probable answers is posed and, as a result of obtaining information I, we are able to select one answer as being certain, i.e. as having a probability of one. Then:

$$I = K \log P \qquad (1)$$

where K is a constant and logs are taken to some convenient base.

This deceptively simple relationship is a cornerstone of information theory, since it allows us to quantify information (although not in the sense of placing values on it, which is a subjective matter) and to investigate the efficiency with which information is transmitted or received during any communication. Information is usually measured in binary digits (bits) calculated from:

$$I = \log_2 P \qquad (2)$$

The equivalence between information and negative entropy leads to the conclusion that the minimum amount of negative entropy lost in obtaining or transmitting a single bit of information is equivalent to approximately 10^{-16} cgs units of entropy, regardless of the method of transmission (Brillouin, 1956). The proportionality is so small that microbiologists do not need to worry about the entropy cost of obtaining information about their systems; nevertheless, its existence is vital to the cell in regulating every one of its physical and chemical processes.

In information theory, information in the form of a coded sequence of *symbols* constituting a *message* is often considered to pass through space from a *transmitter* to a *receiver*. These two entities are connected by a *channel*. In ideal situations the channel behaves perfectly; it is noiseless and of infinite capacity. In many practical situations, however, which are describable by information theory, the channel is noisy, i.e. it introduces an uncertainty as to whether a particular received symbol is actually the one transmitted. The detection and enumeration of microorganisms are examples of the transmission of information, and the relationships concerning the transmission of information, particularly along noisy channels, are definitely of concern to the microbiologist. In microbiology, one may sometimes consider the transmission of information from one point in space to another, as in the visual detection of areas of altered optical density in a growth vessel, or through time, as in the growth of a colony at the site of the original cell (inoculum).

Signals and Noise

It is quite common for a bacteriologist to determine whether or not a species of organism is present in a given quantity of food, usually 10 or 25 g, but occasionally 100 g. Since a bacterial cell may have a mass of 10^{-12}g or less, the analytical method used for some species is sensitive to at least one part of bacteria in 10^{14} parts by weight of food. With one or two dollars worth of materials and a few days of patience, the bacteriologist performs this exercise in detection easily. We wish to design instruments that will provide an answer more rapidly and with less need for manipulation of materials. However, no physical method available today is capable of directly detecting this concentration of bacteria. We should enquire why.

Physical detection methods rely on some property of the bacteria or their products, such as absorption or emission of electromagnetic radiation, radioactivity, ionization current, electrical impedance, etc. None of these properties is entirely specific to the bacteria; always there are other materials with similar properties present which interfere with the sensitivity of the detection system. The microbial cell is a very small *signal*. Those materials in the specimen having similar properties represent *noise*. Consider a general example where the bacteria contribute volume fraction v_B of the specimen and are to be detected by some property X, such that their contribution to the measured value of X is given by:

$$[X_B] = v_B X_B \qquad (3)$$

where X_B is the specific value of X for unit volume of bacteria. There will also be materials $1,2, \ldots i, \ldots n$ present in the specimen, for which, in general:

$$[X_i] = v_i X_i \qquad (4)$$

and

$$\sum_{i=1}^{i=n} v_i + v_B = 1 \qquad (5)$$

The measured value of X will be

$$X = v_B X_B + \sum_{i=1}^{i=n} v_i X_i \qquad (6)$$

This represents a steady background against which, in principle, the bacteria could be detected. However, there will always be small perturbations, $\Delta (v_i X_i)$, in the various X_i due to small changes in temperature, humidity, pH, etc. or in the dimensions of the measuring cell and so on. Thus X may change to:

$$X + \Delta X = v_B X_B + \Delta (v_B X_B) + \sum_{i=1}^{i=n} v_i X_i + \sum_{i=1}^{i=n} \Delta (v_i X_i) \qquad (7)$$

We can ignore the term $\Delta (v_B X_B)$ and therefore we can see that the bacterial signal must be detected against the background fluctuation:

$$\Delta X = \sum_{i=1}^{i=n} \Delta (v_i X_i) \tag{8}$$

In telecommunications, a signal is generally reckoned to be detectable if its rms voltage is at least twice the rms noise voltage. For the microbial detector, it is reasonable to assume that the microbial signal is detectable if $v_B X_B > 2 \sum_{i=1}^{i=n} \bar{\Delta} (v_i X_i)$, the average fluctuation of X during the measurement period. However, since in the limiting analytical case where an organism is detected in 100 g of food:

$$v_B \leqslant 10^{-14} \sum_{i=1}^{i=n} v_i \tag{9}$$

In order for the microbial signal to be detectable, therefore,

$$X_B \geqslant 2.10^{14} \frac{\displaystyle\sum_{i=1}^{i=n} \bar{\Delta} (v_i X_i)}{\displaystyle\sum_{i=1}^{i=n} v_i} \tag{10}$$

and since $\displaystyle\sum_{i=1}^{i=n} v_i \simeq 1$

$$X_B \geqslant 2.10^{14} \sum_{i=1}^{i=n} \bar{\Delta} (v_i X_i) \tag{11}$$

The detection of microorganisms by any property is thus likely to pose an almost insuperable problem of stability, unless the property X is such that $\sum\limits_{i=n}^{i=n} v_i X_i$ is vanishingly small. In prac-

tice, only one property has so far been found to be sufficiently specific that $\sum\limits_{i=n}^{i=n} v_i X_i$ is zero. This is the property of *multipli-*

cation, the basis of most microbiological methods. The property of *mobility* can be almost as specific; however, for many bacteria, X_B for this property is also zero, so that detection of microorganisms by their movement is of limited value. Fluorescent antibody staining is one of the few other methods even remotely approaching this specificity.

As already stated, the detection of microorganisms in a specimen can be described as the transmission of information through a noisy channel. The measured property of the microorganisms can be regarded as message symbols transmitted from the specimen (the transmitter) to a receiver (the microbial detector) along a channel (the detection method, e.g. light absorption). The existence of noise introduces an uncertainty, i.e. a probability p, that a detected symbol was actually the symbol transmitted. The quantity of information received by the detector (I') is then less than the information represented by the bacteria:

$$I' = I - I_{lost} \tag{12}$$

where I_{lost} is the information lost because of the uncertainty. It can be shown (Shannon and Weaver, 1949) that:

$$I_{lost} = -I[p \ log_2 \ p + (1-p) \ log_2 \ (1-p)] \tag{13}$$

Frequently we are unable to determine a value of p and therefore of I_{lost}. However, a common situation is that the microbial signal is below the limits of detectability, in which case we are quite unable to say whether our observed signal is due to bacteria or the background. In such a case, $p = \frac{1}{2}$. This leads us to the result that:

$$I_{lost} = I \tag{14}$$

and the not very surprising conclusion that:

$$I' = 0 \tag{15}$$

i.e. we have obtained no information about the microorganisms in the specimen.

The Microbial Amplifier

An amplification or detection scheme* for microorganisms may be described by the flow diagram shown in Figure 3-1. For the purpose of illustration, light absorption is shown as the basis of the detection method, but the scheme is quite general. The sensitivity of the microbial detector shown in Figure 3-1 is limited by the average value of the sum of unaccountable fluctuations of X (optical density, in this case) in the specimen, in exactly the same way that the sensitivity of an electronic amplifier is limited by the rms noise voltage at its input.

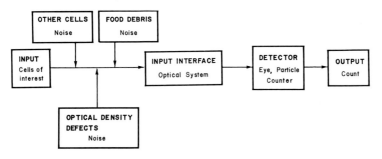

Figure 3-1. Schematic representation of an amplifier or detector, used to detect a microbial property. The specific case where this property is optical density is illustrated. The small microbial signal must be detected against a background of optical noise in the sample.

In most microbiological analyses, however, an incubation stage long enough to allow twenty-five or thirty cell divisions is employed. One can regard any system consisting of one microbial cell, plus sufficient nutrient to allow that cell to grow and divide, as an amplifier. The fact that this amplifier amplifies itself is not particularly important. Each microbial division can be regarded as an amplification stage having a gain of two. Figure 3-1 can, therefore, be expanded as shown in Figure 3-2, to include up to

---

*The terms *amplifier* and *detector* may be used interchangeably.

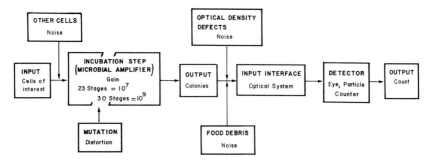

Figure 3-2. When an incubation period is used in a microbiological analysis, the microbial amplifier (see text) acts as a preamplifier to the final detection stage. The microbial amplifier is extremely selective, and each stage has a gain of 2. Noise in the sample is not amplified. The output signal of this amplifier is much more likely to be detected against the background noise.

thirty or more amplification stages, each with a gain of two.

Fortunately for the microbiologist, the microbial amplifier, unlike most physical amplifiers, is virtually distortion-free and extremely selective. Only a living microbial cell (the input signal) can be amplified into two microbial cells (the output signal) and the distortion, represented by the occasional appearance of mutant cells in the output, is probably less than 10^{-9}. All other possible sources, such as fat globules, tissue organelles, or other components, representing amplifiable noise in the simple physical detection scheme of Figure 3-1, are rejected by the microbial amplifier, which is able to provide noise-free amplification factors of 10^9 or more as a result. After an incubation time T, for microorganisms having division time t, the output of this microbial amplifer is approximately $2^{\frac{T}{t}}(v_B X_B)$, a much stronger input for the final (physical) detection stage, and much more likely to be detected reliably against the background noise.

Of course, if a particular species of microorganism is to be detected, microbial noise in the form of interfering contaminant species may also be present. In such a case, the microbiologist may improve the selectivity of the microbial amplifier still further, by modifying the growth medium or other growth conditions.

In a typical example, the analytical method is the pour plate counting technique, and the organisms exist at a level of 10 per g in a specimen of sausage. The final detection stage may be the human eye, aided or not by a microscope, or an electrooptical particle counter. Before incubation, neither detector can detect the small signal represented by the organisms against optical noise such as sausage components, bubbles or ripples in the agar, and optical defects in the Petri dish. At the end of the incubation period, the output signal from each microbial amplifier is a colony containing say, 10^9 bacteria. The light scattering or attenuation corresponding to this output signal is now sufficiently strong for it to be detected by the human eye and probably by the particle counter. The use of this microbial preamplifier gives the microbiologist a tremendous increase in the sensitivity of his detection method, but at the expense of time. Thus, the overall rate of transmission of information is much lower than if a physical detection method could be used directly.

Errors and Redundancy

The existence of noise in a transmission channel introduces errors in the received information, since there is now an uncertainty as to whether a received symbol was actually the symbol sent. It is impossible to transmit information entirely without errors along a noisy channel, but the number of errors introduced may be reduced to any desired finite amount by coding the information in a suitably redundant form, i.e. by adding to a message additional symbols which act as checks on the accuracy of other parts of the message. The simplest form of redundancy is repetition; the more times the message is repeated the greater is the probability that the average value for any particular received symbol correctly indicates the transmitted symbol. This form of redundancy is pertinent to *signal averaging,* whereby a weak signal is slowly recovered from a noisy input. For a message repeated r times:

$$\text{Redundancy } R = 1 - \frac{1}{r} \tag{16}$$

Thus the redundancy may take positive values between 0 (message sent only once) and 1 (message sent an infinite number of times).

In telecommunications simple repetition redundancy is rarely used because it reduces the rate of transmission in inverse proportion to r. In microbiology the duplication of inoculated Petri dishes for each dilution represents a form of repetition redundancy, as does any plurality of data from which a mean is taken. However, there seem to be only a few instances of redundancy being used to improve the accuracy of transmission of microbial information. The preparation of serial dilutions of a sample can be regarded as a means of coding the microbial information in order to minimize the number of errors introduced by noise, such as chance contamination of Petri dishes, or the randomness of the occurrence of microorganisms in the inocula. For example, if decimal dilutions are used in a counting technique, we expect the number of colonies in successive Petri dishes to exhibit an approximately decimal decrease in numbers. The failure of this expected progression to materialize indicates that some form of noise is present, and the manner of the failure may also indicate which of those values we can reasonably trust. The science of the construction of error detecting and error correcting codes is beyond the scope of this chapter. However, it may be that there exist schemes which are intrinsically more suited to mechanized counting methods than our conventional ones.

It is important to note that redundancy only improves the accuracy of transmission of information if the detector is able to make use of the redundant information. Thus, although addition of irrelevant symbols to a message increases the total content of the message and decreases the rate of transmission of information, the accuracy of transmission may then actually decrease since it becomes even less certain whether a particular symbol or sequence of symbols was actually sent. Excess information is as undesirable as, or even indistinguishable from, noise. The case of the electronic particle counter, described later, illustrates this.

Minimum Information Needed From the Analysis

It is interesting to calculate the theoretical minimal quantities of information required in various types of microbiological analyses. The figures obtained can serve as bases from which the efficiency of some of the available methods of detecting microorganisms can be evaluated.

The following examples are relevant to machines that we might reasonably construct to carry out microbiological analyses. It is assumed that a given machine has no prior information about a sample on which it is required to perform an analysis. Consider first the total count of organisms where, in a typical case, the number of organisms in a sample can be assumed to lie between 1 and 10^8. If the machine has no prior information about this number, we can assume that there are 10^8 equally probable answers to the analytical question, from which the machine must select one as being correct on the basis of information supplied by the analytical method it uses. In order for it to be able to specify one answer, the analytical method must supply:

$$I = log_2 \ 10^8 \tag{17}$$

or 26.6 bits of information.

If we were content with having the machine grade the specimen in decimal steps, e.g. Grade A = 1 to 10; Grade B = 11 to 100, etc., the analytical method would only need to provide information sufficient to distinguish between eight grades. We should then preferably redefine the assumption regarding probabilities by saying that all grades are equally probable. Then, the analytical method must supply only:

$$I = log_2 \ 8 \tag{18}$$

or 3 bits of information.

The information provided by counting the colonies on a Petri dish is limited by the maximum number, e.g. 300, of colonies allowed on a dish. Actually, if we only allow the counting of dishes containing between 30 and 300 colonies there are only 270 possible answers, from which the machine must select one as being correct. Thus the method of counting a Petri dish must supply a minimum of:

$$I = log_2 \ 270 \tag{19}$$

or 8.1 bits of information.

If we wish to identify a certain strain or species of microorganism in a specimen, without regard to its enumeration, and if the machine has no prior information about the nature of the organism, the number of possible answers is equal to the number of known strains (x) capable of existing in the specimen. This number is not, of course, fixed, nor in practice are the various answers equally probable. However, to our disinterested instrument all answers may be considered equally probable, and the quantity of information to be supplied by the method is, therefore:

$$I = log_2 x \qquad (20)$$

If we also wish to enumerate a particular species or strain, the quantities of information required for the two operations are additive. If I_e is the information required to enumerate the organism and I_i the information required to identify it, and these values are independent, the total information required to enumerate the strain is:

$$I = I_e + I_i \qquad (21)$$

It often happens that we merely wish to know whether or not a certain strain exists in a specimen. For example, we may wish to know whether a food sample contains a certain strain of *Salmonella* during the tracing of a food poisoning outbreak. In such a case it might be possible to employ a specific fluorescent-antibody stain, in which case there would be only two possible answers (*yes* or *no*), and the quantity of information supplied by the method would be:

$$I = log_2 2 \qquad (22)$$

or 1 bit of information.

In general, microbiological analytical methods represent such noisy channels that it is impossible to obtain sufficient information to allow one answer to be specified with certainty. Uncertainties in the received messages reduce the information content of the method (equation 13), leaving us with a residue of possible answers. This is so even when a computer is used to upgrade the information content of data by means of various pattern recognition techniques, or by comparison with *a priori* possibilities of events.

Electronic Colony Counting

The counting of conventional colonies on the surface of a Petri dish or membrane filter requires examination of the whole growth area in order for the colonies to be detected. Consider the case of a counting circle just filling the vertical frame height of a 525 line rectangular television scan of 625 line horizontal resolving power. In such a case, the circle occupies approximately two thirds of the area of the rectangle, and the number of picture points into which the image of the circle can be divided is thus $\frac{2}{3}$ $\times$ 525 $\times$ 625, or 216,000. For simplicity of comparison with the "most probable number" method, described below, the brightness of the television image may be considered capable of being quantized into 64 levels (symbols) rather than being continuous. The total amount of information carried by a single television frame is therefore:

$$I = log_2 \ 64^{216,000} \qquad (23)$$

or 1,296,000 bits of information.

The quantity of information in the television scan is thus very much greater than is required to enumerate 30 to 300 colonies in the circle (8.1 bits), or even to enumerate the organisms in a food sample (26.6 bits). The communication problem lies in the fact that only a very small proportion of the total encoded message from the scanning of the circle represents sequences of symbols which will later be identified as colonies by the discriminator of the apparatus. There is a great deal of excess information, most of which is unfortunately not in a form where redundancy can help to improve the accuracy of the transmission. It is obvious that the greater the ratio of total to encoded information, the greater will be the probability of spurious countable sequences arising as a result of noise.

The likelihood of the information encoded by the microbial colonies being transmitted at all usefully would be very small if it were not for the inclusion of suitable bandpass filters in the counter. The filter reduces the rate of transmission of information, reducing the noise power level in a manner analogous to signal averaging or repetition redundancy. Obviously, the more uniform in size the colonies are, the narrower can be the pass-band, and the less the effect of noise. Even so, it is unlikely that the information-carrying capacity of the channel can be made to match the information encoded by the colonies, a requisite for accurate transmission. It is probably safe to conclude that counting methods based on scanning of large growth areas and a step-by-step comparison of optical properties, such as are used by existing electronic counters, are plagued by noise simply because of this mismatch between the information-carrying characteristics and the information carried.

Owing to the effect of "nearest neighbor inhibition" techniques, used to avoid registering a plurality of counts from large colonies, it is most important that the sequence of symbols be received in proper sequence during the scan, otherwise the machine can never decode the message so as to obtain a meaningful count. The coding of the information contains no redundant features. The situation is very different when the most probable number methods, such as the MPN tube test, the microtiter plate, or the hydrophobic grid-membrane filter (HGMF) are used as the basis of the analytical method. In these methods, the "colony" occupies the whole of each separate area of examination.

Consider a microtiter plate or HGMF, having a pattern of N growth compartments and scanned in such a way that only the center of each growth compartment is examined for its optical density. In this case, the detector only registers whether or not each growth compartment has reached a certain optical density. Thus there are only two possible levels or symbols (for example, 0 for no growth, 1 for growth) and 2^N possible combinations of symbols. The information coded by the pattern of growth compartments and which could be received by the detector is thus:

$$I = log_2 2^N \qquad (24)$$

or N bits of information.

Typically, N is 96 for the microtiter plate and 1,024 or 10,000 for the HGMF, so that the total information encoded by the pattern of growth compartments is much less than for the randomly inoculated plate, which must be scanned. However, it is obvious that by the time all the 1 values are summed in the counter there are only $(N+1)$ total messages conveyed by the sequences of symbols (0 to N positive growth compartments). The order in which the symbols are received is now unimportant. The actual information content of the message is thus only:

$$I = log_2 (N+1) = log_2 N \quad \text{if } N>>1 \qquad (25)$$

We can write that the redundancy is:

$$R = 1 - \frac{log_2 (N+1)}{N} = log_2 \frac{2N}{(N+1)} \qquad (26)$$

It can be seen that this redundancy tends to unity as N increases, and by the time $N = 1,024$ or 10,000 the redundancy is very high. That this is closely related to repetition redundancy can be seen in the following illustration, in which two 0s and two 1s are combined in six different but equally probable ways, thus:

$$1,1,0,0\ldots1,0,1,0\ldots1,0,0,1\ldots0,1,1,0\ldots0,1,0,1\ldots0,0,1,1$$

The occurrence of any one of these sequences of four symbols would produce a count of two. Unlike the case of examination of a Petri dish or membrane filter for randomly located colonies, the order of the message symbols is unimportant, and this must have an effect on the accuracy with which the information can be transmitted; it is probably the most accurate possible way of coding the microbial information. The overall accuracy is still limited, of course, by optical noise, i.e. by the certainty with which a received symbol can be ascribed to the growth or nongrowth of microorganisms. This depends on the signal to noise ratio of the optical system (*see* Chapter 10) .

A similar effect would be obtained by using an array of N photocells in a pattern corresponding to the pattern of growth compartments, rather than the scanning beam of the television camera, since such a device would merely have $(N+1)$ output states.

The examination of a specimen for the presence or absence of a particular species or strain of organism closely resembles that for the enumeration of colonies with respect to accuracy. Consider, for example, the electronic examination of a microscope

field for cells stained with fluorescent-antibody, compared with a
method has been described by Munson et al. (1970) (*See also*
which a stained culture has been filtered is examined. Such a
method which has been described by Munson et al. (1976) (*See
also* Chapter 8). In the first case, the location of any fluorescing
cell is unknown, and the whole field must be scanned, leading to
the production of a superfluity of mainly useless information, as
has been seen for the electronic colony counter. In contrast, the
examination of a membrane filter for fluorescence can be reduced
to a simple yes/no situation yielding just one bit of information.
The total quantity of information required to enable one to say
whether the organism is present or not is also one bit. One can
regard this experiment, in fact, as a single compartment MPN
method. Again it is likely that the more effective matching of
coded information to the channel makes this method inherently
more accurate than any method employing electronic scanning of
a microscope field.

Mechanical Enumeration of Microorganisms

It is interesting to investigate briefly the conditions required
for enumeration of microorganisms by machine, and to compare
this with what is feasible. Consider a machine using an analytical
technique based on the plate count, i.e. making a series of dilutions of constant factor, and plating these out in such a way that
colonies can be counted. Assume also that the machine is capable
of making such mathematical calculations as are necessary to obtain a result, but is incapable of comparing results with *a priori*
probabilities.

If the sample contains up to n microbial cells, there are n equally probable answers to the analytical question. The quantity of
information required by the machine to enable it to specify one
particular answer is:

$$I = log_2 n \qquad (27)$$

Imagine that the machine carries out an analysis using d growth
vessels, each capable of resolving m colonies. The machine makes
$(d\text{-}1)$ dilution operations, each of dilution factor D, such that:

$$n = mD^{(d-1)} \qquad (28)$$

In the analytical method itself, there are md different and equally probable results (the maximum number of possible answers that can be obtained from d growth vessels each capable of having m colonies). As a result of carrying out the analysis, the machine is able to pick one of these to use in calculating n. Therefore, in the absence of noise, the analysis yields a quantity of information:

$$I' = log_2\ m + log_2\ d \qquad (29)$$

If the analytical method allowed the transmission and reception of the microbial information without loss, I' would equal I, so that:

$$log_2\ m + log_2\ d = log_2\ m + (d-1)\ log_2\ D \qquad (30)$$

or,

$$log_2\ d = (d-1)\ log_2\ D \qquad (31)$$

Values satisfying both equation 31 and condition 28 are given by:

$$d = \frac{n}{m} \qquad (32)$$

and

$$D = \left(\frac{n}{m}\right)^{\frac{m}{n-m}} \qquad (33)$$

There are a maximum of n possible ways ($m = 1$ to n) of formulating the analysis so as to obtain $I' = I$, i.e. in order to permit enumeration of the microorganisms. Actually there would generally be fewer, since d should be an integer. At the two extremes we have:

i. $m = 1$; $d = n$; $D = n^{\frac{1}{n-1}}$; i.e. the machine uses n growth vessels each capable of resolving one colony, and a dilution factor of $n^{\frac{1}{n-1}}$

ii. $m = n$; $d = 1$; $D = 1$; i.e. the machine uses one growth vessel capable of resolving n colonies, and does not make dilutions.

Either way, to specify n (to enumerate the microorganisms) would require the use of very inconvenient numbers or sizes of dishes or dilutions in the machine if n is of the order of 10^8. It should be noted at this point that our conventional manual plate count does not precisely enumerate microorganisms either. If we prepare six plates for the sample, and agree to count plates containing between 30 and 300 colonies, then our analytical method only provides sufficient information for us to choose from 6×270, or 1620 different answers. Thus, if $n = 10^8$, there are 99,998,380 answers that we cannot specify. We can conclude that it is not feasible to construct a machine that will enumerate microorganisms precisely, and that we must be content with merely grading the sample as we do in the conventional plate count.

If we approach the problem positively and consider constructing a machine to *grade* samples, the feasibility is quite different. Consider, for example, a set of grades based on exponentials of some constant b. The number of grades into which the sample can fall is thus $log_b n$. If we assume that all grades are equally probable, the amount of information needed to allow us to specify in which grade the sample belongs is:

$$I = log_2 log_b n \qquad (34)$$

Imagine again that the machine prepares d growth vessels and $(d\text{-}1)$ dilutions of constant factor D, such than $n = mD^{d-1}$. Since we only wish for the machine to say whether a growth vessel contains growth or not, we may write, for the moment, $m = 1$. The experiment then yields a quantity of information:

$$I' = log_2 (d{-}1) \qquad (35)$$

If the information encoded by the microorganisms is transmitted and received without loss, $I' = I$, so that:

$$log_2 (d{-}1) = log_2 log_b n \qquad (36)$$

from which we see that:

$$d = 1 + log_b n \qquad (37)$$

and

$$D = b \qquad (38)$$

The logarithmic form of equation 37 leads us to a much more feasible state of affairs. The grades can obviously be made as close as desired by using a sufficient number of dilutions, although a compromise must eventually be reached between the closeness of the grades and the engineering feasibility of preparing the corresponding number of dilutions.

In practice, because of the normal distribution of microorganisms in the liquid inocula, there will only be a probability *(p)* of 0.645 that a given volume of inoculum containing an average of one microbial cell will actually yield a colony of growth. Thus, a certain amount of information (described by equation 13) will generally be lost if the grading machine prepares only one growth vessel for each dilution. It is obviously preferable, therefore, either (a) to prepare from each dilution a growth vessel capable of holding a plurality of colonies (in which case grading is based on whether or not the number of colonies in the vessel is above or below a certain figure) or (b) to inoculate a plurality of compartments similar to MPN tubes, microtiter plate wells, or HGMF grid-cells (in which case grading is based on whether or not a certain fraction of the compartments contains growth). Bearing in mind what has already been discussed regarding the relative accuracies of detecting colonies or altered growth compartments *(see* **Electronic Colony Counting**), it would appear that the latter method would be preferable. The confidence with which the sample can be placed within a particular grade can be made as high as desired by increasing the number of compartments inoculated from the corresponding dilution.

From the above argument, therefore, it can be suggested that if we wish to construct machines that will carry out analytical procedures yielding answers in terms of numbers of colonies or growth forming units, we may do well to consider the following format: The machine will be described as a grading machine. It will prepare serial dilutions of the sample having a factor corresponding to the width of the grades. It will prepare from each dilution a plurality of discrete growth compartments inoculated with volumes of the dilution, the number of compartments being determined by the level of confidence with which the sample must be graded. After incubation, it will examine each growth compartment individually for growth, on a yes/no basis, and will classify the sample as being above or below each grade level according to whether or not a predetermined fraction of the compartments for each grade is positive.

REFERENCES

Brillouin, L.: 1956. *Science and Information Theory.* Acad Pr, New York.

Goodyear, C.C.: 1971. *Signals and Information.* Butterworths, London.

Hancock, J.C.: 1961. *An Introduction to the Principles of Communication Theory.* McGraw, New York.

Munson, T.E., J.P. Schrade, N.B. Bisciello, L.D. Fantasia, W.H. Hartung, and J.J. O'Connor: 1976. Evaluation of an automated fluorescent antibody procedure for detection of *Salmonella* in foods and feeds. *Appl Environ Microbiol, 31:*514.

Pierce, J.R.: 1961. *Symbols, Signals and Noise: the Nature and Process of Communication.* Har-Row, New York.

Rosie, A.M.: 1966. *Information and Communication Theory.* Blackie and Son, London.

Shannon, C.E. and W. Weaver: 1949. *The Mathematical Theory of Communication.* U Ill Pr, Urbana.

Young, J.F.: 1971. *Information Theory.* Wiley, New York.

Chapter 4

CONSIDERATIONS AND PROBLEMS IN DEVELOPING AN INSTRUMENT FROM CONCEPT TO COMMERCIAL PRODUCT

D. FREEDMAN

Introduction

SINCE the early 1940s, the field of microbiology has gone through a most remarkable transformation. With the discovery and subsequent use of penicillin, man began to appreciate more fully the potential of microorganisms for useful purposes. Urgent demands for penicillin throughout World War II thrust microbiologists, biochemists, and chemical engineers into an urgent program to develop processes and instruments in areas which were in many ways unfamiliar to them. Immediately following the war, industrial fermentation was rapidly developed to an advanced stage, to a degree that microorganisms are now used to produce a host of complex chemicals, antibiotics, enzymes, and vitamins, performing highly specific tasks to change complex chemical molecules.

During this period, fresh impetus was given to the design and development of the technique and instrumentation required for control and manipulation of microorganisms. Many scientists throughout the world embarked on attempts to fabricate the equipment and instrumentation necessary to harness and control the environment in order to get new microbial products and by-products.

There are numerous inventors of microbiological and biochemical instruments who have made the advancements in this field possible. At the same time there are scientists who have spent inestimable amounts of time developing instruments and techniques and yet have not seen their developments result in commercially acceptable products. It is this area of taking a concept or idea through its development to a product that is considered in this chapter.

Initial Considerations and Problems

The scientist's needs are the compelling force for the development of most laboratory instruments and apparatus. As is obvious in most fields, the very best and largest number of ideas for microbiological instruments are generated within the microbiology laboratory by the workers within these facilities. No one can know the problems as well as the person who has to deal with them on a daily, and even hourly basis. Although some of the New Brunswick Scientific products are cited, the emphasis of this discussion will be directed towards the methods and steps that most manufacturing companies pursue in their product development programs.

The most obvious stimulus for the development of a new product is the suggestion for a tool to fulfill an existing need. Throughout the years, scientists have expended an inordinate amount of time in developing tools and apparatus which, while needed, distracted effort from the main research programs.

One important aspect of new products development is the conviction by a company that there is a market for a particular product. The "need" for a product may manifest itself in many ways, such as time saved from routine operations within the laboratory, improved methods within the laboratory, or greater applicability of an instrument because of increased accuracy. Very frequently, the presence of a highly accurate instrument becomes the stimulus for doing research in a particular direction. When an instrument company believes that by producing a new product new levels of research will probably be pursued, this additionally becomes an incentive for new product development. Quality control regulations and purity standards are also stimuli for new products, as are process requirements which individual companies or industries may generate. This certainly takes place in the pharmaceutical field, where companies develop their own tools to match their process development requirements. Some of these tools go on to become general laboratory apparatus.

Many marvelous ideas generated within laboratories are, unfortunately, never picked up by companies and manufactured. There are also situations where products are developed within a research laboratory and then redeveloped ten and twenty years later when the value of the concept is finally recognized. Circumstances dictate the manner in which instrument manufacturing companies and organizations vary their techniques and methods of approaching new product development. The five-man instrument company obviously cannot carry out the same type of development program as a pharmaceutical company with a staff of 5,000. There is, however, a group of small to moderate size instrument companies that have a common approach to marketing, designing, and manufacturing instruments.

When consideration is given to the development of a product, the first step taken is a preliminary marketing study. Is the product concept and design acceptable to the market? Are customers' needs presently being satisfied? What is the size, location, and make-up of the market? Does the market, for example, include only microbiologists, or other specialists as well? Is it limited to research laboratories, or does it include clinics and hospitals?

At the same time, the organization wants to estimate the potential for market growth. Depending on the type of item, a study is made of the competition and the potential for competition. An evaluation of the probable price range in relation to preliminary design features is undertaken, as well as of the distribution channels required. Is this a product which lends itself primarily to direct sales or one which should be sold on a very broad basis and therefore most suitable to dealer distribution or wide direct sales distribution? Obviously, brand preferences in relation to competetive or similar equipment must be explored. To ascertain whether an idea is worthwhile pursuing, the opinions of users are sought, including salesmen. To provoke greater interest, a set of preliminary specifications is developed, primarily to establish a model or a concept which is complete enough to obtain necessary marketing information. This generally requires putting together some preliminary engineering information to give the idea some form or shape.

Assuming that preliminary information is encouraging, a model is then manufactured. When this is done, the organization is in a position to establish a market or sales survey by one or more methods. A specification sheet and photograph of the equipment can be prepared and survey questionnaires sent out, either directly or through independent organizations, to a cross section of the potential market segments. Then a market-by-market evaluation of the survey is arranged. Also helpful is a telephone survey for simpler types of products; a telephone inquiry is also frequently useful following a mailed questionnaire.

More useful than this, however, is arranging for scientists who work in the particular field to gather to discuss the product.

Another very valuable approach is a direct personal interview with potential users using a questionnaire and, if possible, a preliminary model. This invariably stimulates ideas and constructive criticism which adds substantial understanding to the needs of the laboratory. The demonstration of a preliminary model in conjunction with a questionnaire at a trade exhibition is also most helpful. This may stimulate groups of scientists to volunteer suggestions and new ideas, or alternative approaches and applications.

Those companies which engage consultants or scientific advisors outside their firm can usually arrange private surveys, which can be more meaningful and thorough because of the time commitments involved.

After all the preliminary marketing data is assembled and responses validated, a management review of the results takes place. Design specifications and price are reappraised in light of the data, and then a second generation of preliminary specifications is developed. This includes all the pertinent technical information. Some market estimate, indicating probable high and low limits within a specified time frame, is made. Also, a preliminary sales price is established as a guide for engineering and design considerations, since engineering will be done with a view toward manufacturing at a specific cost level. A preliminary marketing plan and budget must be established; this would incorporate an applications plan.

If the product involves a new concept or complex procedure, or if communication is difficult, then a User's Guide or an Application Guide should be prepared. This tool will help clarify problems involved in the design. It will also help explain the procedures and will assist in establishing the features which should be designed into the equipment.

Also, at this early juncture an evaluation of Food & Drug Administration requirements for the product should start. Regulations on Medical Devices and Diagnostic Products are presently being developed and are likely to increase in the United States as well as in other countries.

For more information about compliance regulations, contact:

> Food & Drug Administration
> 8757 Georgia Avenue
> HFK-123
> Silver Spring, Maryland   20910

Developmental Engineering

We now enter the engineering phase, although various parts of the procedure may actually be going on simultaneously. There are normally six engineering phases required in the development of a complex product for manufacture. Each phase has different goals and requires different talents and expertise. The first phase is feasibility; the second development; the third, design; the fourth, pilot manufacture; the fifth, a testing program; and the sixth, final manufacturing and performance specifications, alteration and correction of drawings, and final engineering release. Also, part of the engineering plan is designation of a person who will be responsible for establishing final specifications between the engineering and marketing groups.

The first phase, establishment of feasibility by scientific analysis and/or experimental verification, is usually accomplished by reviewing the various concepts involved. The result is a more polished set of technical specifications and refinement of cost/time estimates. The technical specifications vary according to the type of equipment to be developed. The specifications found in sales literature are only a fraction of the engineering specification generally established during this phase of engineering. Here, for example, we would want to work out the process time, perhaps methods of heating, power efficiency, and the heat transfer involved. For example, if the device is a simple water bath, a requirement to heat from 20°C to 60°C in a particular amount of time might be approached in a variety of ways. One way of transferring heat rapidly into the fluid could be efficient, while another could be quite inefficient. This sort of thing would be included in the general specifications, but not those used for sale literature. As an example, considerations are frequently given to keep the wattage level of an apparatus moderate, so as not to overload circuits, but still to give a reasonably effective performance.

As another example, the specifications for the (New Brunswick Scientific Company's) Automatic Electronic Colony Counter® cover an extremely large number of aspects, including a general description, general requirements, electrical power requirements and tolerances, ambient temperature limits, operating temperatures, nonoperating temperature, humidity range tolerances, warm-up time, physical characteristics, a listing of what the standard unit will include (such as a nine-inch video monitor, top and bottom light, video reversal system, size control, electronic blanking, mechanical adjustment mask, stage, lens sizes etc.), the outputs and optional accessories to be included, mechanical and sheet metal design specifications, light specifications, optional specifications, video signal specifications, image quality specifications, stability specifications, contrast level specifications, circularity tolerances, and, finally, general standards. Of course, this only provides a specification outline, but it is close to the final specification.

Having gone through the feasibility phase, the next phase is a study of the various techniques suitable for the implementation of the concept. This may require the development of new devices to be used in conjunction with presently available components and usually involves extensive study and experimentation before the optimum techniques are selected. Various experimental assemblies are usually constructed and tested during the development phase to prove that the desired performance can be achieved. This testing should not be confused with the testing that occurs in the fifth phase of the engineering program.

The third phase is conversion of the experimental equipment into a design that can be manufactured with known or specially developed methods. During this phase, consideration is given to components, circuit and component reliability, and the check-out of various systems that appear to be required. Layout and engineering drawings are made, the esthetics of the device are reviewed, and industrial design concepts are considered. An industrial designer may be involved in both the appearance and the human engineering of the product. Here also consideration is given to equipment size and ease of operation, maintenance, and repair. Since the initial model, previously used for market survey purposes, may not in fact be the final configuration, a series of artist's renderings may now be drawn and paper or wood models made to permit further evaluation of design features and appearance.

Also during the design phase, all detail drawings, components, and materials to be used are selected, and a total bill of materials is produced. The bill is detailed into parts, subassemblies, and then final assemblies, and outlined in an indented fashion to show how one assembly goes into the next and the latter into the next, and so on. It covers all of the various models of a particular product, as distinct from a parts list, which might only deal with one specific model. A replacement parts review also takes place (for design considerations). Performance specifications become further detailed, and inspection specifications are created. One of the important facets of this phase is the generation of specification control. This allows an inspection (or quality control) department to be certain that all purchased parts meet the specifications on which the design was originally based. It is quite common for component manufacturers to change the specifications of an item without changing its description or model number, thereby causing the end product to function in a manner different from the way it functioned during the engineering phase.

Pilot manufacture (phase four) takes place after the design

phase is completed. Generally, pilot manufacture results in the production of from one to many hundred units, depending on the particular product. At New Brunswick Scientific, if the design review appears to be satisfactory, three prototypes are produced. These are made without any special tooling and are usually referred to as "handmade" models. These models include all of the purchased components which have been designed into the product. Since tooling is frequently unavailable when models are produced, substitute materials may be used. Various companies, particularly those in the clinical field, will sometimes use these models for field testing after they have been tested in their own laboratory. In these cases, the companies would usually produce a considerable number of pilot models.

Phase five, the testing program, is one of the most important. In this program, tests will be run under what are considered "normal operating conditions," as well as under abnormal conditions. For example, if a piece of equipment requires water for its normal operation, it should be tested without water to determine the extent to which it is affected by such improper use. The unit would also be tested with waters of different quality and temperature, to predict results in different parts of the country and world. When steam is to be used, the engineer would check not only with clean steam but also with steam contaminated with particulate matter, to see whether the dirt will affect the system. The ambient temperature would obviously be one of the factors to consider, and it would also be desirable to determine, for example, the possible adverse effects of sunlight shining on the equipment through a window. The engineer would test the electrical systems under low voltage conditions and check for things such as motor overheating and the effect of heat produced by the motor etc. on the other components. Humidity limitations or requirements, nonoperating temperatures, and abrasion during operation would be checked. A program to simulate shipping conditions would also be investigated. Sterility testing programs would be established where applicable. Obviously, all of the safety features would be carefully examined to insure their proper function. The equipment would be tested with all of the accessories that were designed for use with it. It would be cleaned as might be anticipated under actual use, for example, in instances where washing, scrubbing, and splashing might take place, these conditions would be duplicated. The equipment should be used in every possible way that the user would be expected to employ, and it ought to be misoperated in every conceivable way. Unfortunately, every possible misoperation is not considered; users are frequently far more ingenious.

Pilot models may be sent out for field trial. Also, at least one unit will be set up for life testing on a continuous operating cycle basis, to determine what may fail and when failures can be anticipated. There is always something that fails, and it is best to find which components or weaknesses will be the most troublesome.

After all of the testing data are analyzed and design corrections examined, the product goes into a final engineering phase (phase six) where both manufacturing and performance specifications are completed, all of the engineering information is reviewed, the drawings checked, and those areas which may have been found to be troublesome corrected. The designs for different electrical systems to cover various parts of the world are included in this final phase.

It is interesting to note that because of the various electrical systems in use throughout the world, there is a great lack of compatability between electrical components and terminology. For example, in the United States, 220 volts means 240 volts to some people, and 208 volts to others. In the United Kingdom 220 volts is usually 240 to 250 volts. In the United States, 440 volts usually is 480 volts.

Also in this final engineering stage, a shipping container is designed, final operating instructions are written and the bill of materials is finalized. At this point, a complete set of information and blueprints is made available for the production departments. A blueprint is essentially a means of providing clear information to the manufacturer of the product or its parts. In some companies, the bill of material is part of the blueprint, while in others it is a separate series of sheets which can be used for various functions, including purchasing, production control, engineering, and fabrication. Inspection or quality control information is provided and generally reviewed with the relevant department, and a review of tooling requirements is prepared with the pertinent manufacturing department. The engineering department provides all known purchasing information to the purchasing department. This will include names of vendors, as well as alternate sources. The purchasing group will then review the product with the manufacturing engineering department, which is responsible for methods and tooling, and with the production control people, who are responsible for the production planning.

Manufacturing Engineering and Production Control

After engineering has formally released all its information, a major procedural alteration goes into force. If any change is required in any portion of the product, a written engineering change notice must now be generated. These engineering change notices are distributed to all departments involved with the product, such as inspection, purchasing, planning, and the various manufacturing branches, in order to control the change desired.

Subsequent to the release of the completed design information, the manufacturing engineering department begins to play a critical role. It establishes the methods by which the varying components are to be manufactured, and what tooling is required. Decisions are based on the projected number of items to be produced. Tooling will naturally vary with the number of items to be manufactured, also with the type of equipment the company possesses. Manufacturing engineering will establish in-process inspection requirements for all fixtures, and the inspection tooling. They will also set up operation sheets for the manufacture of parts, the tooling for subassemblies, and the tooling and methods for the final assembly. The production control and planning functions take over after manufacturing engineering is complete, although many of these functions proceed simultaneously.

The production control group reviews the bill of materials and separates out the parts list to facilitate ordering. They establish production schedules based on orders or sales projections and create inventory control methods to maintain data relevant to purchased materials and manufactured parts. Whether the procedure is completely manual or computerized, or a combination of the two, the object of the production control function is to have all of the required materials available at the time each phase of production takes place, but to have it on hand for the shortest possible time. The production control group, in conjunction with manufacturing engineering, also establishes man-hour requirements in the various production departments and generates the machine-loading requirements (how many parts can be made on a given machine in a given time period) and the man-loading requirements (the number of parts a particular person can produce in a given time). This data will be used to coordinate the availability of materials and manpower. Production control also arranges and coordinates the movement of parts and subassemblies from the manufacturing departments, vendors, subcontractors, and warehouses.

The purchasing department purchases the required materials, either in raw form or as components, and attempts to keep control of costs and find superior sources or improved components. As indicated earlier, however, changes of vendors or components must not take place without the use of an engineering change notice.

Generally, in small companies, the inspection department and the quality control department are one and the same. In larger companies, these functions are separated. Some form of inspection usually takes place during all phases of manufacturing. At New Brunswick Scientific, all parts are inspected, whether they are manufactured internally or purchased. Even the most minor screw is inspected to determine, for example, whether the proper item was received. Of course, for some items, sampling procedures are established so that not every single part has to be handled. Generally, however, individual parts are inspected, then the subassemblies in which these parts are used are inspected, and then the final assembly is inspected. If the material goes through a finishing operation, for example painting or plating, that aspect is inspected too. The final assembly involves functional testing, at which point the blueprints and operating instructions are tied together to see that everything is coordinated.

For new items, a pilot manufacturing run is usually initiated before a longer manufacturing run is made. The quality control department will generally take at least one unit from this first production run for additional life testing. Records are kept of all the inspection phases and testing procedures. After inspection, the product is then moved to the packing area from where it is shipped to the customer or to a warehouse.

Patents

At various times during the development of a new product the question of secrecy and patent production may come up for consideration. Today, most scientists, engineers, and inventors, as well as marketing and advertising people, are quite conscious of patents, trademarks, copyrights, and methods of protecting their original concepts. Large companies staff departments to deal with these matters, as do most of the major universities and research institutions. Nevertheless, extensive misinformation about these subjects is demonstrated by the manner in which industrial and commercial institutions misuse the applicable patent terminology.

In the United States, patent attorneys are usually required to have a law degree and one or more engineering degrees. This is also true in the Western European countries. As a layman in patent matters, the author is only able to discuss the subject as it applies to the United States, since he has limited exposure to foreign patent conditions.

Many scientists believe that by cancelling an envelope containing invention material at the post office one can establish this as a date for patentability of a new invention. This is not true in the United States and is generally untrue throughout most of the world. What is necessary is that an appropriate application be filed with the Patent Office for the particular protection desired.

The first United States patent law was passed April 10, 1790. The power to grant patents at that time was placed with the Secretary of State, the Secretary of War, and the Attorney-General. Thomas Jefferson, then Secretary of State, personally examined all patent applications. The law was revised in 1793, and in 1836 a new law was established which included the examination system of granting patents. The law was rewritten in 1870, and again in 1952, under Title 35 — "Patents" of the United States Code.

The administration of the law concerning granting of patents and related activities has been placed in the United States Patent Office, a bureau of the Department of Commerce. The head of this office is the Commissioner of Patents. About half of the employees are examiners and others with technical and legal training. The United States Patent Office is a very large organization and one of the most outstanding sources of information covering almost any technology imaginable. It is probably the most thorough area for patent examination in the world and is considered by many to be the most difficult office from which to obtain a patent. Approximately 60 percent of the patent applications filed here are granted. A patent may be granted for the invention of any new and useful process, machine, manufacture or composition of matter, or any new, useful improvement. The key word in this simple definition is *new*.

One of the first requirements for patent approval is that the patent application be submitted to the Patent Office within one year of the invention. Also, the invention must be new and original; no prior submission of information or publications covering the same device can have taken place anywhere in the world. It is further required that the concept or idea would not have been obvious to a person having ordinary skill in the art. The inventor has one year prior to filing in which to publish, allow public use of, or manufacture and sell the invented product. The usual time required in the United States for issuance of a patent is a year and a half to three years after the initial submission, depending on the complexity of the case. The life of a patent in the United States is seventeen years after issue.

Application for a patent can only be made by the inventor or inventors of the device. This is sometimes very nebulous since when groups of people are involved in the generation of a new product the ideas come from many sources. The best ideas may come from the engineers or, on the other hand, from the microbiologists. Disputes concerning this matter can lead to a lack of validity of a patent. To avoid such potential problems, it is good practice to maintain very meticulous records when a new idea or concept is under consideration and development.

A patent is an item of personal property and it is only issued to an individual or a group of individuals. It may be sold to others, mortgaged, bequeathed by will, or passed to the heirs of a deceased inventor. A patent may also be assigned to another person, organization, company, etc., by means of a written agreement, either after it is granted or while it is in the application stage. A person working for a company, a university, or a research institution frequently "signs away" the rights to a patent as a prerequisite of his employment. Therefore, although an individual may be the inventor and have a patent issued in his name, the patent is almost always owned by his employer.

A patent usually consists of an abstract of the disclosure, background of the inventions, a summary of the invention, drawings and a description of the drawings, a description of the preferred method of design, a list of claims, and a bibliography of the previous art. The key to the patent, however, is the claims; they are the essence of what is covered by the patent and are the legal aspects in question when litigation occurs.

An important requirement when filing is an oath of inventorship, which is submitted with the appropriate filing fee and remaining patent documents.

When applying for a patent, it is not necessary to describe the exact product or item. In fact, some applicants describe a different approach than that which they intend to produce. Their claims cover their invention without revealing the real concepts involved. This is very characteristic of large companies and organizations, who feel this to be the preferred method. In order to strengthen one's patent position, it is often desirable to obtain multiple patents on a product.

After a patent application has been filed with the Patent Office, the inventor or the organization involved is allowed to use the references "Patent Applied For," "Patent Pending," or "Pat. Pend.". Although inventors may use these references, the applicant never knows whether a patent will be granted. Furthermore, other manufacturers may copy the invention during this period without any recourse by the inventor. Legal recourse is only available after the patent has issued. There are also other conditions in the United States that allow an infringer to copy a patented product without recourse on the part of the owner. Issuance of a patent can be accelerated under certain conditions of infringement, and, at the present time, if the item being patented has a bearing on environment protection.

When a patented product is sold without patent number affixed, the patent is usually considered dormant. Thus, just having a patent is not sufficient. The manufacturer has the responsibility of informing the public that the product is covered by patent. He usually does this by affixing conspicuous labels.

There are many different types of patent law firms to assist inventors. These firms may deal exclusively in patent law or handle patents as part of a general law practice. The latter is more common in foreign countries. It is unusual for a practicing attorney to become involved in patent litigation, since very specialized expertise is normally required; he would generally refer litigation to specialized firms. Patent litigations are usually extremely expensive. If a suit goes through to a final judgment, the cost could be thirty to fifty thousand dollars for a simple litigation, and very much more for a complex one. This is one of the reasons why product licensing is such a common practice.

All patent information is freely available and can be reviewed by attorneys or by the general public. The complete patent file is usually referred to as the "Patent Wrapper."

Patent litigation in the United States takes place within the federal court system. Decisions are made by judges who are appointed to the Federal District Courts and who generally do not have great expertise in judging the validity of patents. Usually they have had no technical training, apart from on-the-job training. Therefore, court actions usually require the legal firms to present very elaborate, educational, and technical material in order to present their case in a clear and well-defined manner.

The success of upholding the validity of litigated patents is surprisingly low, between 22 and 55 percent in various federal districts. The Federal District Court in the New York region has the lowest record (about 22%). More recently the statistics are improving in favor of the patent holder.

In the United States, a patent owner has a right to sue for infringement with a minimum of exposure. The court may award the patentee damages to compensate for the infringement (which may include the profits made by the infringer) and has the power to increase the damages up to three times the amount found. The court can also grant an injunction to prevent continuation of the infringement. Defendants usually challenge the patent validity in their defense. Both parties normally pay their own legal fees and court costs. Foreign patents usually favor the owner of the patent, and the foreign courts usually award a winner-take-all decision. The losing party pays the legal expenses for both sides, as well as some level of restitution and penalty.

It might be noted that when there is financial aid by a government or agency of government, such as the National Science Foundation, an invention is generally considered public domain. Therefore, any invention which may be made by an individual working with public funds becomes suspect if the invention is in the field covered by those funds.

In litigation, if the defendant's activities do not fall within the language of any of the patent claims, they will ordinarily not be held to infringe. In the United States the losing party in the suit can appeal to the District Court of Appeals but not to the Supreme Court. The federal government may use any patent invention without the consent of the patentee, but the patentee has the right to compensation through negotiations. If no agreement can be reached, the patentee may sue the government for restitution in the Court of Claims.

Many firms maintain secrecy, rather than patenting their inventions, to avoid having competitors find a way of circumventing their ideas. Some laymen, as well as scientists, in their enthusiasm for royalties, sometimes put too much weight on patent protection. Royalties generally vary from 1 to 5 percent of the net selling price. A strong agreement with a modest royalty may be more valuable than a large royalty with a weak agreement. Most companies will honor patents and license them if the royalties are moderate, but tend to look for alternative approaches if the royalty demands are excessive.

Fees

When filing for a United States patent, there are several fees required: an initial filing fee of approximately $65.00 to $100.00, as well as an issue fee of $130.00 to $150.00. Small amendment fees are also charged, as pertinent. Additionally, the inventor or his assignee will ordinarily pay fees for attorneys, draftsmen, proofreaders, and perhaps consultants. In the United States, unlike every other country in the world, once you have a patent no additional payments are required. The cost for most simple apparatus patents (Utility Patents) generally involve a total cost in the vicinity of $1,000.00 to $3,000.00.

Foreign Patents

The major countries of the world have reciprocal agreements which allow patentees time to file in other countries after the initial filing. The most important patent treaty is one called "The International Convention for the Protection of Industrial Property" (International Convention), to which more than sixty countries belonged by the mid-1960s. The "Convention" provides one year from the date of the original filing for patents to be filed in other countries. If one files outside the United States within this one-year period, the United States filing date is the effective date in foreign countries; if one files after the one-year period, the foreign date is the effective date. Although the "International Convention" protects the original filing date, it should be carefully noted that each country has its own requirements for prior art and public use. The countries considered most important for patents in the scientific instrument field are the United States, Canada, the United Kingdom, the Netherlands, Germany, France, Switzerland, and Japan. Patent law of the Soviet Union provides authors of inventors with certificates of patent only, which provides for some payment but not for patent ownership.

The rule in the United States requires that an application be filed no more than one year after the invention is described in a printed United States or foreign publication, or be in public use or on sale in the United States. Thus, a printed publication disclosing the invention is effective regardless of where it is published, while "public use" or "on sale" requires that the activity take place in the United States. In many foreign countries, the publication, "public use" or "on sale" condition prior to filing invalidates the patent. Some countries include the whole world as their jurisdiction, while others include only their own country. Under some of these circumstances it is quite difficult to obtain a valid patent in a foreign country, even though the patent is valid in United States; the inventor is not likely to find this out until the patent becomes involved in litigation.

Foreign filing fees are usually lower than those for the original patent, because the applications are essentially copies of the original. The translations, however, do add to the total cost. Unlike the United States, all foreign patents have a life of fifteen to twenty years and are taxed annually at a continuously increasing rate. For example, the following is the current fee schedule in West Germany:

Year	Fee
2	$ ―――
3	44.00
4	44.00
5	66.00
6	99.00
7	132.00
8	176.00
9	220.00
10	264.00
11	352.00
12	462.00
13	572.00
14	682.00
15	792.00
16	924.00

17	1,056.00
18	1,188.00

Foreign patents therefore, become very expensive to maintain, especially when old.

Design Patents

A Design Patent is generally subject to the same rules as a regular patent for an invention (which is sometimes referred to as a Utility Patent). It is issued for a term of three-and-a-half, seven, or fourteen years, in accordance with the applicant's choice and payment of the corresponding fee. An application must be filed in the name of the inventor or inventors, although the resulting patent is usually assigned to a company or organization. A Design Patent generally covers the visible external characteristics and ornamental configuration of an item. It relates only to appearance, and not to construction or structural characteristics. Design Patents are not widely used for scientific equipment but are used frequently for items such as toys and games, of wide commercial circulation. Patentability is predicated on the novelty of the configuration or surface ornamentation. A design which merely simulates a known object is generally not deemed patentable.

The elements of applications for Design Patents are similar to those for Utility Patents, but the description is generally confined to the drawing, and only a single claim is made to the design as shown. The cost of patent preparation and filing is generally a small fraction of that for Utility Patents. "Patent Pending" is also applied to Design Patents. To indicate this, the prefix (D) should be affixed to the item and incorporated in the packaging of promotional material involved.

Foreign filing of Design Patent applications varies according to the country. Generally speaking, the right to priority with respect to filing is six months, as distinct from one year for Utility Patents. It is desirable to file a Design Patent application as soon as the external configuration has been finalized. The application must be placed on file within one year of the first public use.

Trademarks

Virtually every company that manufacturers equipment or consumer products uses trademarks to a great extent. A trademark can consist of a brand name, such as "Gyrotory," "MicroFerm," "Dixie" (as in Dixie cup), "Sanka;" or a slogan, such as "His Master's Voice" (for RCA), "All the News That's Fit to Print" (for The New York Times) ; or a design, such as the NBS Erlenmeyer Flask, "I.H." (for International Harvester) ; or any other combination of words or symbols which serve as a indication of general identification.

The best trademarks identify the product either by a symbol or words, without describing the product. Trademarks which sound too much like an actual activity or apparatus are much less satisfactory, since they tend to be descriptive of the goods, and are therefore, very difficult to protect.

Once a trademark has been selected, it is desirable to check trade directories or any other source which might disclose anything pertinent to that trademark. The next step ought to be a search in the United States Patent Office, which has a listing of all trademarks registered in Washington. This search would cover both registered and pending applications. It should be noted, however, that there are many unregistered trademarks in use. If heavy expenditures in advertising are anticipated, a thorough search of all trademark sources, although it may be fairly expensive, is desirable.

To obtain federal registration for a trademark from the United States Patent Office, it is necessary that a bonafide commercial shipment of goods bearing the trademark be made across state lines. The invoice or other record of such interstate shipment should be maintained in the organization's files. A record of the first local shipment should be maintained as well. As soon as a product bearing a new trademark is shipped across a state line that trademark has been acquired, and trademark rights exist at that point. It is now necessary to obtain registration and to indicate next to the trademark that trademark rights are being claimed. This is usually done by affixing the letters "T.M." or the word "Trademark" immediately following the trademark.

Application for registration is filed at the United States Patent Office. The examiner will determine whether there is any conflict with the registration. If there is not, it will be published in the *Patent Office Official Gazette.* Those considering that they might be adversely affected by the registration can oppose the application or request an extension of time to do so. An opposition proceeding can tie up an application for a number of years. If the application is not rejected and there is no opposition, the registration will issue fairly quickly. Once a federal registration has been obtained from the Patent Office, the ® notation should be used immediately after the trademark. This symbol can only be used after a mark has been registered.

In order to maintain registration, a document must be filed between the fifth and sixth year after registration, indicating that the trademark is still in use. If this is not done, the registered trademark will be cancelled automatically. The trademark will expire after twenty years, unless the registration is again renewed at that time. This must take place within six months prior to the end of the first or any subsequent twenty-year period. Registration can be renewed indefinitely, providing the trademark remains in use.

The rules and requirements of other countries differ from those in United States. However, it is possible to register the trademark in most foreign countries, based upon the United States registration, even when use has not taken place in the foreign country. Foreign trademark registrations are usually recommended in countries where there are strong commercial interests.

The best and most immediate way of protecting your trademark in the United States is to use the trademark or brand name upon or in association with the goods in commerce. It should always be prominently displayed. It should not be used in the plural form and should not be confused with words indicating the style, type, or model of the equipment. Registration of a trademark may be cancelled if the trademark is not used for a period of two years or more.

Copyrights

Copyrights may be obtained for items such as books, publications, advertisements, labels, catalogs, price lists, operating manuals, sheet music, phonograph records, maps, works of art, reproductions of works of art, ornamental packaging, drawings, photographs, and motion picture and television films. To protect a new invention, it is usually advantageous to copyright the pertinent literature, sales brochures, advertising materials, operating instructions, and published specifications.

A common law copyright automatically exists for all works prior to distribution such that, if it is wrongfully taken or used in advance of distribution, suit can be brought against the infringer. However, copyright generally refers to statutory copyrights. To obtain this, the work involved must bear the copyright notice when distributed, an application for registration of claim to copyright must be filed with the Register of Copyrights in Washington, and a certificate issued. The most important prerequisite is that the copyright notice appear on every copy of the work. If registration is obtained, and at a later time the copyright notice is omitted from distributed material, all protection is lost.

A copyright notice for publication by New Brunswick Scientific in 1976, would appear as follows: "Copyright © 1976 New Brunswick Scientific Co., Inc.". If there are space limitations, the symbol © can be used without the word "Copyright", but in that case the © must be very clear, not smudged or obliterated. Abbreviated names may also be copyrighted and marked with the name and the Copyright ©. Where the work involved is written material, such as a booklet, pamphlet, or brochure, copyright notice with the year must appear prominently on the title page or on the first page of the text. Copyright notice in the center of the booklet or on the back page or back cover is not valid.

Copyright infringement can consist of outright copying or sufficient similarity between the works involved that it is apparent that the second was inspired by the first. Copyright litigation is complex but, unlike the laws relating to patents or trademarks, the successful plaintiff in an infringement suit can obtain not only an injunction restraining further distribution, but also damages sometimes calculated at one dollar for each copy distributed and reimbursement of legal fees.

Copyrights presently exist in the United States for a period of twenty-eight years from the date of first publication and can be renewed for an additional twenty-eight year period if the application for renewal is filed before the expiration of the original term. The maximum term for such protection in the United States is a total of fifty-six years. The law applying to United States copyrights is currently being reviewed by Congress and changes will probably be made.

Marketing

The pilot production run is usually the key to establishing the final selling price. This run will provide an actual cost and actual manufacturing history. By the time the run is underway, marketing plans are usually well established and sales literature has already been produced and circulated.

At this stage, it is desirable to have a rather complete guide to the application of the new product. If handled well, such a guide becomes one of the most valuable tools in the whole marketing operation. The key elements of the product will be carefully discussed, on a scientific basis, so that any scientist would be able to tell from the guide if the instrument or device will be applicable to his or her work.

The placing of a completed production model in appropriate laboratories for testing and use is most valuable. It provides an excellent opportunity to determine where there may be problems and flaws, and even what may be the best method of marketing. By this time, an advertising and promotional program will have been started. This will include exhibits at scientific meetings, technical seminars on product use and performance, the dissemination of press releases and articles to technical and trade journals, publication of technical literature and supplementary material, and the introduction of the products to potential customers by a sales force. By producing sales aids, such as photographs, case history reports, and audio visual material and charts, one can further supplement the sales program. Journal advertising and direct mail, if handled properly, will usually bring in a substantial number of inquiries. Additional salesman training and further indoctrination is required, as well as preparation of customer education and training manuals. In order to evaluate these activities, the results of each effort should be evaluated, and fed back by the sales force. Conducting workshops in various places may assist the marketing effort. It is also of great value to have one's own applications laboratory personnel use the equipment or instrument on a regular basis. Consultants are also helpful because they may assist by using the equipment on a long-term basis, providing detailed information on a regular basis.

Since the final selling price is likely to be established after the initial pilot run, the efficiency of production during this period must be considered in relation to that expected during subsequent manufacturing. Amortization of engineering and development costs is usually a component of the final price; amortization of scientific equipment normally covers a three-year period. Among the key factors to market growth is customer satisfaction, and feedback on this is, of course, crucial. After-sales service, therefore, becomes an important factor.

Most companies will establish a selling price based on what they think will be accepted, even if it is beyond a "normal" markup. On the other hand, many companies will accept less than their normal profit margin in order to accelerate market penetrations and establish themselves firmly with the particular product. Initiating sales on a less than normal profit basis also discourages competitors, thereby giving the particular company a headstart in terms of product identification. A large number of companies, however, use some type of fixed formula to establish their price on the basis of actual or estimated cost.

Advice to Inventors

For the microbiologist who is inventive and innovative, with design capability, some avenues are available for converting ideas into completed products. However, to get the product satisfactorily manufactured and effectively marketed, the proper firm with expertise in the particular specialty must be selected.

The product line and markets in which the manufacturing company is involved should be reviewed. A rejection of the idea or a lack of interest should not lead to discouragement.

Many companies make mistakes in evaluating new products. One merely has to look in virtually any field of endeavor to confirm this point. The inventor should not be discouraged, even after repeated failure to achieve acceptance of an idea. If the same organization is contacted more than once, the inventor will frequently deal with different people. It is probably worthwhile to reintroduce the same product to the same company several times with new information to reinforce your initial presentation, since the thinking of a company will often change. Many executives learn very slowly about new products, especially when it is a matter of grasping a new concept.

If a product idea has a large mass market, the larger companies with extensive worldwide distribution facilities that can blanket the marketplace quickly ought to be approached. The large corporations, however, are sometimes more difficult to deal with and require a good deal more persistence. In presenting an idea, the inventor should provide comprehensive information concerning the description and operation of the equipment, the market served, the advantages of the product, and any scientific references, articles and publications, market data, and studies that can facilitate the manufacturer's evaluation.

Time is one of the very important elements to control in negotiating with an instrument company, since the evaluation could drag along endlessly unless there are some time limitations. In working out an agreement one should establish modest royalties but also establish a minimum guarantee in terms of volume commitments. The person to be contacted usually should be the most progressive-thinking person accessible within a company. This is not necessarily high-level management. The person contacted, however, should have a major influence in reaching a corporate decision. Sales and marketing managers frequently fall into this category.

There are endless opportunities for the development of new ideas, products, and services in the field of science. Most of the

companies in the scientific field have always been actively interested in the examination and pursuit of any useful product, and
inventors can almost always find a warm reception for most new
ideas.

Chapter 5

STATUS OF MECHANIZED MICROBIOLOGICAL STANDARD METHODS FOR FOODS

R. B. READ, JR., J. E. CAMPBELL, AND J. S. WINBUSH

Abstract

To the detriment of microbiology, development of instruments for use in official automated methods has been slow. The authors believe that this is due, in part, to the small potential market and the cost of development, including the cost of manufacturing, supplying, and servicing several instruments for use in collaborative study—all without any certainty that the instrument will withstand collaborative study and be marketable. Since the potential market is fixed, little can be done to improve this aspect. However, the design of collaborative studies could be modified and with this, the need for a manufacturer to supply several instruments for study purposes. Statistical techniques are available that can be used to evaluate studies involving only a few instruments. These techniques are being assessed for use in collaborative studies as recommended by the Association of Official Analytical Chemists.

MICROBIOLOGICAL standards for raw milk and for some finished dairy products have been in use in several countries for decades, and efforts are being made to develop international standards for these products. Recently, there has been renewed interest in microbiological standards for foods other than dairy products, and standards have been either proposed or adopted for some foods in many countries. The authors believe that the adoption of microbiological standards for foods will grow both nationally and internationally for use in purchasing specifications and for use as one of several regulatory tools directed toward improvement of food safety and quality.

For standards to be useful they must be based upon sampling techniques that are adequate to obtain a representative portion of the food being tested, and the methods of analysis must be readily reproducible so that results obtained both within a given laboratory and among many laboratories are comparable. Practical realization of these goals is accomplished in the United States by using methods that are listed as "Official" by the Association of Official Analytical Chemists (AOAC) or as "Standard" in "Standard Methods for the Examination of Dairy Products" published by the American Public Health Association (APHA) because they have been shown to give results that are reproducible in many laboratories. In the succeeding discussion, these methods will be referred to as standard methods.

When the AOAC and the APHA publications are examined, it becomes immediately apparent that there are no standard microbiological methods that can be characterized as mechanized or automated. Furthermore, as far as the authors are aware, collaborative studies are presently in progress on only two mechanized methods: a study of the several electronic colony counters for counting bacterial colonies in Petri dishes and the spiral plating procedure for bacteria in foods developed in the Food and Drug Administration. Since collaborative study is one of the prerequisites for a standard method, it appears that there will not be many mechanized standard microbiological methods in the immediate future.

Because chemical examination of foods is a common regulatory tool, it is tempting to compare the mechanization of chemical regulatory techniques with their counterparts in microbiology. Analytical chemistry has developed to a point where instrumental methods of analysis appear to be the rule rather than the exception. The marked contrast between this situation and methods for regulatory microbiology immediately provokes the question as to why the difference exists. In chemistry, two basic reasons for mechanization seem to predominate — lower limit of detection, and decreased cost of analysis. In microbiology, lower levels of detectability do not appear to be a viable reason for mechanization. As an example, methods used by the authors for *Salmonella* in oysters have been shown to detect three or four *Salmonella* cells in 100 g of oyster meat. If the presence of one *Salmonella* cell in 25 g of oysters is assumed and assuming that a *Salmonella* cell has a diameter of 0.5μ, the detectability of the method expressed in chemical terms is about 16 parts per quadrillion or 16×10^{-15}. This level is four to five orders of magnitude lower than many of our most sophisticated analytical chemical techniques.

One of the characteristics of a few of the mechanized microbiological methods that have been developed is that daily start-up time is longer for the mechanized method than for the conventional technique because of reagent preparation and instrument calibration. This disadvantage is readily overcome through the greater speed with which samples can be tested once the mechanized procedure is in operation. However, the combination of faster analyses coupled with increased start-up time makes mechanization attractive only if there is an appreciable sample volume. Unfortunately, at present few microbiological laboratories have large numbers of samples available at any one time in the laboratory day that need the same kind of analyses. This lack of sample volume, coupled with no improvement in detectability from the mechanized method, are believed to be rather basic inhibitory factors in the development and application of mechanized microbiological methods.

Presently, microbiological testing is used by many segments of the food industry for determining whether purchases meet specifications and for within-plant quality control purposes. The application of microbiological standards will undoubtedly result in the initiation of new quality control procedures in some food manufacturing and warehousing firms, as well as the broadening of quality control procedures in other firms in the food industry. If this does occur, the volume of testing will increase and thus provide a greater opportunity for mechanized methods.

The authors have participated in many meetings at the request of prospective developers of mechanized methods applicable to food microbiology. Typically, the developer would describe an idea in general terms and then ask basically four questions: (a) What is the size of the potential market in food microbiology? (b)

What foods are tested most? (c) What is the upper limit of mechanized instrumentation cost that could be justified? (d) How does one get a method approved as standard?

In spite of the authors general belief that mechanization would be beneficial to regulatory microbiology, they have felt compelled to give the following answers: (a) The potential market is relatively small in the U.S.; (b) The most tested food is probably milk; (c) From experience, only a few United States regulatory laboratories will buy expensive instrumentation for microbiology; and (d) Before a method is official or standard in the United States, it must withstand collaborative study to show that it can give reproducible results in routine analytical laboratories, and usually the developer will be asked to supply the instrumentation needed for the study of the proposed mechanized standard method.

This rather discouraging characterization of the immediate potential for mechanized methods in regulatory microbiology certainly does not stimulate developers to go ahead, which is unfortunate for microbiology. Obviously, any opinions such as these offer great opportunity for error, particularly in the absence of experience with mechanized microbiological methods for foods. However, experience in the use of mechanized methods in related areas outside of microbiology may be useful in evaluating the immediate potential for such methods in regulatory microbiology and the required instrumentation. As an example, milk is the food most often tested microbiologically for regulatory purposes. Current regulations in the United States require that raw milk be tested for somatic cells as often as it is tested microbiologically. Somatic cells in milk have been traditionally tested by microscopic counting of stained milk smears. Two mechanized methods have been developed in recent years and recognized as standard. Both of these involve electronic counting of somatic cells in suspension, replacing the need for the microscope. Although these are not microbiological methods, they are performed typically by the same people who perform microbiological analyses on milk. Of the 800 laboratories in the United States that test milk for regulatory purposes, fewer than thirty are using either mechanized method. This is particularly striking when one considers that they are offered as replacements for an especially tedious procedure — counting cells through a microscope.

In food microbiology we have been and will probably continue to be committed to counting bacterial bodies — generally after a substantial period of time has elapsed so that multiplication makes it easy to perform a count. Many attempts have been made to circumvent this counting, generally through measurement of a metabolic end product or a cell constituent, e.g. methylene blue reduction or analyses for ATP. Unfortunately, metabolic products and cellular constituents vary among species and this variation prevents a high correlation with the plate count. We seem to continue to expect that a replacement for the plate count will correlate well with a plate count. Thus it appears, at least for the immediate future, that any mechanized method proposed for determining bacterial density must count bacterial bodies. In counting bacteria on plates, as an example, considerable judgment is required to obtain reproducible results. Since mechanized techniques cannot use human judgment, it is difficult to develop automated methods that give results that compare precisely with those from conventional procedures.

The development of mechanized methods is usually associated with a series of instrument modifications and improvements particularly in the early years of marketing an instrument. These modifications and "improvements" create a problem when the method that uses this instrumentation has been collaboratively studied and approved as a standard method. At some point in method and instrument modification, one must declare the new method to be different enough from that studied collaboratively to require a new collaborative study. Obviously, the requirement for a new study, after significant instrument modifications, may inhibit modification and consequently slow the improvement of a mechanized method.

Of the series of problems that appear to inhibit mechanized method development, namely potential market size, laboratory volume, and requirement that regulatory methods withstand collaborative studies, only the latter is amenable to immediate change. However, when the impact of regulatory action based upon laboratory analyses is evaluated, it does not appear desirable to do anything that might have the effect of reducing the precision of laboratory work. Because of this the authors feel that the general requirement of method validation through collaborative study should be required for mechanized methods.

One of the basic tools for the evaluation of a method is the accuracy and precision of the method when it is performed in several laboratories. These parameters are measured during collaborative study, which requires a minimum of five collaborators for a reasonable estimate. Inter– and intra–laboratory variations can be determined because the samples that are sent to the participating laboratories are identical. Unfortunately, instrument costs for most mechanized methods are substantial, and the cost of supplying and maintaining several instruments during a collaborative study may be prohibitive.

There are fundamental differences in collaborative tests involving mechanized methods compared to those involving manual procedures. One of these is the ease with which repeated measurements can be made. From the standpoint of designing a collaborative study, the requirement of several replicate determinations on a sample does not impose the additional workload that it would if a manual procedure were being used. Consequently, this, and the problem of instrument cost for collaborative study, invite a reconsideration of collaborative study design.

Collaborative studies might be designed using techniques of incomplete blocks where, depending upon the instrument being evaluated, the product, and other circumstances of the procedure, certain interaction effects may be chosen to be confounded on purpose, if information is on hand to indicate that such interactions are unimportant. Depending upon what assumptions are made and what variables are to be estimated, as few as two instruments could be used for a collaborative study wherein both instruments would be placed in each of two laboratories. In the first instance, when the analyses are completed on the split samples, the pair of instruments would then be installed in another laboratory, additional samples would be analyzed and the process repeated. In the second instance, after the analysis is completed on the split sample in each laboratory, one instrument would be set up in each of two additional laboratories and the process repeated until the required number of laboratories had completed the study. Of course, if more instruments were available, one could substantially reduce instrument cost at the expense of additional sample preparation and the overall time required to complete a collaborative study.

In summation, in spite of the rather discouraging assessment of the present and near-future use of mechanized methods in regulatory microbiology, the authors believe that mechanization must be studied and introduced into routine laboratory analyses as rapidly as possible. Interest in mechanized microbiological methods is definitely on the increase, as is the need for mechanization in regulatory microbiology as well as in many other facets of the science. The authors believe and hope that the combination of increased interest and increased need will enable us to make progress at an increasing rate as the science of mechanized microbiology grows from its infancy to maturity.

Chapter 6

EVALUATION OF AUTOMATIC BACTERIOLOGICAL ANALYSIS METHODS APPLICABLE TO TOTAL BACTERIAL COUNTS IN MILK

R. GRAPPIN

Abstract

Increasingly, methods for counting aerobic mesophilic bacteria (total bacterial count) are being used to assess the bacteriological quality of milk. For laboratories that must process several thousand samples daily, the use of mechanized equipment especially designed for this purpose is absolutely essential. There are now several machines for preparing Petri dishes or roll tubes and for counting colonies on Petri dishes. Before the use of such equipment can be given official approval, tests must be done to assess the main characteristics of the apparatus in question. The basic purpose of the tests is to assess the economic value (reliability, performance, cost per sample) and the analytic value of the equipment or methods concerned (fidelity and precision in relation to reference methods). Special attention should be given to calibration and means of checking the results obtained. From measurements made during the tests, conclusions can be drawn as to the potential usefulness of the equipment under particular conditions. An equally important issue is the preparation of standard procedures. The objective here is not to give an exact description of the analytical method, as for manual methods, but to indicate the range within which the instrument may be used and standards for the degree of fidelity and precision that the user can and should achieve with the instrument. These various aspects are illustrated by examples such as the mechanization of a simplified method for determining total bacterial counts and various mechanized Petri dish colony counters used by laboratories in France.

Introduction

AT THE present time, the method of counting aerobic mesophilic bacteria or "total bacterial count" is very widely used to assess the bacteriological quality of food products, particularly milk when delivered by the producer to the dairy. This method is gradually replacing the dye reduction methods (methylene blue, resazurin), which are less suitable for milk that has been kept for several days at 4°C, as is now being done in farm tanks. These tests are not sensitive enough for grading milk with less than 500,000 bacteria per ml, let alone 100,000 bacteria per ml. Manual count methods, however, take time and are hardly practical in control laboratories where sometimes several thousand samples must be analysed daily. Methods using automatic equipment would obviously be far more convenient, since laboratories could do the analyses faster and at lower cost.

Automatic Equipment for Making Total Bacterial Counts of Milk

As several days of incubation are necessary to make the counts, the only steps that can be mechanized are the preparation of culture dishes or tubes and colony counts. Although several types of automatic apparatus exist, relatively few are used regularly by milk analysis laboratories to determine total bacterial counts. For the preparation of Petri dishes or roll tubes, existing automatic equipment is based on the conventional dilution method or on a simplified method. Conventional method is used by two machines: the Colworth 2000® (Sharpe et al., 1972), which uses Petri dishes and prepares all dilutions up to 10^{-8}, and a Dutch machine (Jaartsveld and Swinkels, 1974), which is especially designed for milk analysis and uses roll tubes. Machines based on simplified counting methods include one described by Bradshaw et al. (1973) in the United States, and one by Posthumus et al. (1974) in the Netherlands, which use either a cylinder or a calibrated loop for taking milk samples. The technique applied by the Dutch machine, now in use at the Zutphen laboratory, is very similar to the method described by Thompson et al. (1960) which uses a calibrated loop; the only difference is that the former employs roll tubes instead of plates. Finally, Gilchrist et al. (1973), in the United States have described a simple automatic technique for spiral inoculation of Petri dishes.

To the author's knowledge, only two machines, the Eramic 25® (Grappin, 1975) developed in France and in use there since 1974 and the Petri-Foss® (Foss-Electric, Doc. No. Foss Trading, Denmark), provide complete and accurate mechanical applications of the Thompson method, which is one of the official French methods used to determine the price paid for milk on the basis of the milk's bacteriological quality. In these machines, all the operations involved in preparing Petri dishes are mechanized: the taking of a 1 μl (Eramic 25) or a 10 μl (Petri-Foss) sample by means of a loop, inoculation, addition of agar medium, agitation, identification, and cooling.

To complete the mechanization process, there are a relatively large number of machines, fairly well known for the most part, that can be used to count colonies on Petri dishes. These range from very sophisticated instruments that can analyse images (OMNICON Image Analyser®, Quantimet 720 P Automatic Petri Dish Analyser®) to simpler machines especially designed for Petri dish colony counts. The latter include three that are used in France, the Biomatic® and the Numérateur II® (Jeunet et al., 1973) which analyse the image of a Petri dish obtained from a television camera, and the GP 73®, which uses a photoelectric method similar to a microscopic count. The technique developed by Tolle et al. (1968) and used in Germany should also be mentioned here; it consists of conducting a microcolony count with a Coulter Counter, after culture in a gel medium and incubation under conditions that differ from those used in the conventional method (two days at 20°C instead of three days at 30°C).

Another approach to the mechanization of counting methods, which is often adopted, consists in looking for techniques that are based on the measurement of biochemical characteristics or certain properties of bacteria; such methods lend themselves more readily to mechanization. This approach was used by Tolle et al.

(1972) in developing a method based in measurement of pyruvate in milk. Here we have an indirect approach analogous to the old dye reduction method mentioned previously, which has itself been mechanized to a certain extent.

With the development of various kinds of mechanized equipment, a number of problems have arisen in the area of approval and standardization. Every country has a more or less extensive array of official national methods of analysis for use in quality control of food products, including milk. However, these are almost always manual methods involving basic principles which are sometimes totally different from those of mechanized methods. Even where the principle is the same, there are differences in performance and different causes of error. The same applies to official standards, which involve particular methods of analysis; usually only manual methods are specified. The various national authorities responsible for these matters are now faced with the difficult problem of approval of mechanized methods for use in standards. A thorough experimental analysis is essential.

Evaluation of Mechanized Equipment

The task of evaluating analytical equipment involves a study of three distinct aspects: the validity of the principle on which the equipment is based, its technical and economic value, and its analytical capability. This applies, of course, to any method of analysis.

VALIDITY OF THE BASIC PRINCIPLE. Briefly, does the method provide the required information about the quality or composition of the product to be analysed? Is it valid for the purpose? These questions arise, for instance, when comparing the pyruvate determination method with total bacterial count methods. Can the former give a valid indication of the number of bacteria contained in the milk or does it more correctly measure the metabolic activity of the bacteria, which is not necessarily related to cell numbers? Does it yield an accurate picture of the hygiene observed by the milk producer during milking? The author will not go into the respective advantages and drawbacks of the two methods, but his opinion is that the bacterial count method seems better adapted to the purpose of the analysis, which is to determine the price paid to milk producers on the basis of bacteriological quality.

TECHNICAL AND ECONOMIC VALUE. This depends on a number of aspects well known to all students of the subject, such as reliability, ease of utilization, labor-saving qualities, risks, cost per analysis, capital investment, and so forth.

ANALYTIC VALUE OR PRECISION. Here is where the experimenter's actual work begins, and this aspect will be discussed in some detail. Because of the increasing quantity of mechanized apparatus being used for milk analysis, and also the considerable number of studies being conducted in various milk-producing countries, a draft standard is being developed by the International Organization for Standardization (ISO) to serve as a guide for evaluating mechanized methods. Its recommendations are valid not only for the dairy sector (chemical composition and bacteriological quality) but for all areas of bacteriological analysis. Since assessment of mechanized methods is an important international matter, the main definitions relating to the issues of instrument precision and methods of testing mechanized apparatus will be discussed. In the example, an instrument used for preparing Petri dishes (the Eramic 25) and also some mechanized colony counters will be analysed. It should be understood that the author is concerned here only with the evaluation of quantitative methods; qualitative methods require a different interpretation.

GENERAL PRECISION OF A DETERMINATION METHOD. When a measurement (m) is made with a given method, that measurement is an estimate of the true value (μ) of the quantity present. The precision of the method increases as (m) minus (μ) becomes smaller. The true value μ is usually not known, but in most cases is taken to be the value given by a selected reference method. Basically, the difference (m $-$ μ) depends on: the fidelity of the method, which is inversely proportional to the number of random errors that occur when it is used and the fit, which is inversely proportional to the systematic errors that occur.

FIDELITY. According to ISO standards, fidelity expresses the degree of similarity of results obtained on the same sample during a series of determinations (either carried out under similar conditions, same operator, same apparatus, at very short intervals, etc), in which case we speak of *repeatability,* or carried out under different conditions, particularly in different laboratories, in which case we speak of *reproducibility.* Both repeatability and reproducibility should be examined during inter-laboratory testing, with each laboratory analysing two sets of one or more samples at different concentrations. From the results, we obtain the inter-laboratory variance, which gives an estimate of the repeatability variance (σ^2_r) ; the reproducibility variance (σ^2_R) is obtained by adding the repeatability variance (σ^2_r) and the inter-laboratory variance (σ^2_L) :

$$\sigma^2_R = \sigma^2_r + \sigma^2_L$$

In order to be more precise, and to have a more readily usable numerical datum we define repeatability (r) and reproducibility (R) as being, respectively, the discrepancy in absolute terms between two determinations, x1 and x2, performed under repeatable conditions and under reproducible conditions, when the probability that the resulting figure will not be exceeded is 95 percent. Working from the law of distribution of deviations between the extremes of a distribution, we obtain the following: $r = 2.77\sigma_r$ and $R = 2.77 - \sigma_R$. Repeatability and reproducibility are important aspects of any analytical method, and the user must know these in order to interpret the value of an analysis result correctly.

It should be pointed out that these definitions and this vocabulary were developed basically for manual methods. In the case of apparatus for mechanized analysis, determining repeatability is not a problem, but the same does not apply to reproducibility, which depends on the degree of accuracy with which the instrument has been calibrated. For machines such as the Eramic 25, which do not require calibration by the user, reproducibility is an important characteristic to determine. For other equipment, however, reproducibility will depend more on the calibrations of the various machines used.

Theoretically, a mechanized apparatus should have a repeatability as good as, if not better than, the same method done manually. In fact, many operations involved in an analysis (test sample, dilutions, etc.) are subject to far less fluctuation when done by machine than by hand. For instance, when a milk sample is taken with a calibrated loop, the sample volume depends partly on the size of the loop, but to a much greater extent on how the loop is dipped and withdrawn, particularly how fast it is taken out. It has been demonstrated (Grappin 1975) that the weight of the milk sample may vary by as much as 100 percent, according to whether the loop is lifted out very fast or taken out slowly. When using a manual method, the operator will not always be able to work at the same speed, and two operators will not necessarily follow identical procedures, a fact which may lead to a considerable difference. This can be best avoided by using an automatic apparatus with a constant withdrawal speed; the same withdrawal speed should be used for all apparatus. Mechanical sampling methods thus ensure greater fidelity.

FIT. This is the degree to which the true value, and the average of several results obtained with the instrument (to eliminate repeatability errors) coincide. Since the true value is usually taken to be that obtained by the reference method (the dilution method for the total count of bacteria, for instance) the concept of fit applies only to indirect methods. The difference between the true value and the measured value may be broken down into two factors, one related to the instrument's calibration, and the other to the difference in basic principle between the measured value and the value obtained by the reference method (Fig. 6-1) . We see two important characteristics:

(a) Calibration accuracy expresses the degree to which the observed calibration line of an instrument and the theoretical calibration line coincide. The observed calibration line is defined as the regression line calculated on the basis of a large number of samples analysed by the reference method (x) and instrument (y) .* The theoretical calibration line

---

*The author has followed the custom of taking the reference method as independent method (x) and the routine method as dependent method (y) . The reverse (x as routine method and y as reference method) , however, would be more sensible statistically.

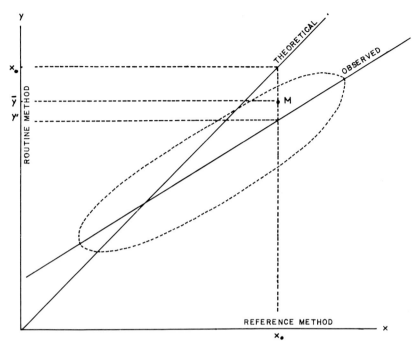

Figure 6-1. Breakdown of main factors determining the fit of a mechanical analytical method.

$\overline{y}$: mean of several determinations, made by the machine, of sample M

x_0 : mean of several determinations made using the reference method

y' : mean value given by the machine for all samples corresponding to x_0

$$\underset{\text{Fit}}{\overline{y} - x_0} = \underset{\substack{\text{Precision in relation} \\ \text{to reference method}}}{\overline{y} - y'} + \underset{\substack{\text{Calibration} \\ \text{accuracy}}}{y' - x_0}$$

corresponds to a line having a slope of unity, and passing through the origin (equation y = x).

(b) Precision in relation to the reference method, which we will call estimate precision, expresses the degree to which the results from the instrument and the reference method coincide, after elimination of repeatability errors and deviations due to any defect in the calibration of the instrument. It is estimated by the distribution of points (residual standard deviation) around the y/x regression line.

The example provided by a colony counter (Fig. 6-2) illustrates these two concepts. First, the regression line indicates a 20 percent underestimate in counts made with the instrument. This underestimate can, in the present case, be corrected by recalibrating. Second, the distribution of points gives an excellent indication of the precision of counts made with the instrument.

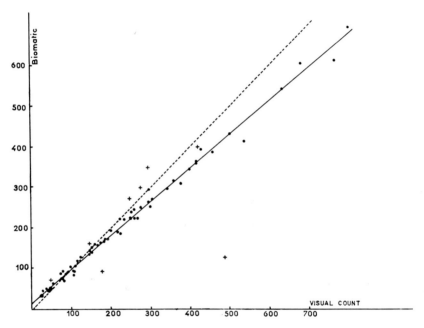

Figure 6-2. Comparison of Petri dish colony counts made with a mechanized apparatus (Biomatic) and by counting visually. The slope of the calibration line could be regulated.

   - - - - - representation of perfect count

  ———— Biometric versus visual line of best fit; correlation coefficient is 0.99

   o Values used to calculate line of best fit

   + Values from abnormal dishes (spreaders etc.) not used in calculations

After elimination of abnormal dishes containing colonies that are star-shaped, spread, stringy etc., the correlation coefficient equally 0.99.

If we discard abnormal dishes, we obtain a standard deviation that varies from 6 when the number of colonies is between 30 and 100, to 25 when the number of colonies is between 300 and 700. In the authors' opinion, the best way to express the precision of a method, in relation to a reference method, is to give the amplitude of variation in 95 percent of the cases, that is, $\pm$ two residual standard deviations (Table 6-I) .

TABLE 6-I

FRENCH STANDARDS FOR REPEATABILITY AND PRECISION
IN RELATION TO THE REFERENCE METHOD OF
MECHANIZED PETRI DISH COLONY COUNTERS
(Number of Colonies)

LEVEL	Size of discrepancy not to be exceeded in at most 95% of the cases	
	Repeatability (discrepancy between two counts)	*Estimate precision (discrepancy in relation to visual count)*
30 - 100	15	$\pm$30
100 - 300	30	$\pm$50
300 - 700	40	$\pm$70

Calibration accuracy in a mechanized instrument is only important if there is no regulating device enabling the user to calibrate his equipment properly. However, precision in relation to the reference method is in all cases a significant measure of the value of an indirect method and must be clearly established by the experimenter.

Briefly, we can say that the analytic value of a mechanized instrument depends basically on its repeatability, precision in relation to the reference method where indirect methods are concerned, on its reproducibility, and on its accuracy of calibration. Examination of an instrument that merely mechanizes or automates a manual method, without modifying its characteristics (Colworth 2000, Eramic 25) does not present any basic problem. It is only important to know the repeatability and possibly the reproducibility and the accuracy of calibration. On the other hand, if the machine is based on a different method, e.g. pyruvate determination and colony counters, not only repeatability but also precision in relation to the reference method must be studied.

Standard Procedure

Since carrying out an analysis with a mechanical instrument is often just a matter of operating the machine and watching the control devices, a standard would seem to be unnecessary. However, although we can hardly expect to set standard procedures for an instrumental method in the same way as is done for manual methods, the user must be given a certain number of instructions in order to avoid major errors. Taking as our example the French standard procedure established for colony counters, we see the following.

Area and Range of Application. Some automatic colony counters suitable for pure culture studies are not suitable for total bacterial counts in milk. Counts made on authorized equipment should be restricted to between 30 and 700 colonies per plate, and dishes with abnormal colonies or very small ones that are not counted by the instrument *should not be counted.*

When an instrument is provided with a device for regulating calibration, a description of the method for checking the instrument's accuracy is required.

Repeatability and Precision: The tolerances allowed for these two criteria are given (Table 6-I).

Generally speaking, mechanization of analytical methods allows both improved repeatability and better reproducibility of results. Regarding apparatus for indirect methods, however, mechanization almost inevitably reduces the theoretical precision of the analyses, e.g. automatic colony counters. Nonetheless, such errors are often negligible compared to errors due to faulty sampling techniques or improper storage; the significance of such errors should not be underestimated.

Because of the technical and economic importance of mechanized methods, parameters indicating their analytic value should not only be well defined but also estimated according to exact standards. The advantages are as follows:

(a) It will be easier to compare the work done by different laboratories responsible for examining new methods of analysis.

(b) Users can be given the main analytic characteristics of the methods in clear and precise language, and in a usable form.

(c) It will become easier to establish standards of precision for the analysis of particular products, for the guidance of manufacturers of analytical equipment.

Again, it must be remembered that of all the criteria considered in this paper, only fidelity is defined by standards. As there is no standard, as such, covering the notion of the fit, the definitions and methods given here should be considered as interim tools only; improvements and refinements may be expected in the future.

REFERENCES

Bradshaw, J.G., D.W. Francis, J.T. Peeler, J.E. Leslie, R.M. Twedt, and R.B. Read: 1973. Mechanical preparation of pour plates for viable bacterial counts of milk samples. *J Dairy Sci, 56:*1011.

Gilchrist, J.E., J.E. Campbell, C.B. Donnely, J.T. Peeler, and J.M. Delaney: 1973. Spiral plate method for bacterial determination. *Appl Microbiol, 25:*244.

Grappin, R.: 1975. Mise au point sur les appareils automatiques utilisés pour la numération des germes totaux du lait : préparation des boîtes de Petri et comptage des colonies. *Rev Lait Françoise, 335:*629.

Jaartsveild, F.H.J. and R. Swinkels. 1974. A mechanized roll tube method for the estimation of the bacterial count of milk. *Neth Milk Dairy J, 28:*93.

Jeunet, R., R. Grappin, M.T. Thiollieres, et J. Richard: 1973. Premiers essais d'un appareil automatique de comptage des colonies sur boîtes de Petri. *Rev Lait Françoice,* 647.

Posthumus, G., C.J. Klijn, and J.J. Giesen: 1974. A mechanized loop method for the total count of bacteria in refrigerated suppliers' milk. *Neth Milk Dairy J, 28:*79.

Sharpe, A.N., D.A. Biggs, and R.J. Oliver: 1972. Machine for automatic bacteriological pour plate preparation. *Appl Microbiol, 24:*70.

Thompson, D.I., C.B. Donnely, and L.A. Black: 1960. A plate loop method for determining viable counts of raw milk. *J Milk Food Technol, 23:*167.

Tolle, A., H. Zeidler, and W. Heeschen: 1968. Die Electronische Mikrokoloniezalung ein Verfahren zur Beurteilung der bakteriologischhygienischen Qualitat der Rohmilk. *Milchwissenschaft, 23*:68.

Tolle, A., W. Heeschen, H. Wernery, J. Reichmuth, and G. Suhren: 1972. Die Pyruvatbestimmung ein neur Weg zur Messung der Bakteriologischen Wertigkert von Milch. *Milchwissenschaft, 27*:343.

Chapter 7

A RAPID METHOD OF ESTIMATING TOTAL BACTERIAL COUNTS IN GROUND BEEF*

L. R. BROWN AND G. W. CHILDERS

Abstract

A market survey on ground beef samples conducted in 1955 showed a close correlation between the oxidation-reduction potential (Eh) and the total bacterial count of the sample. A subsequent study under controlled conditions was conducted on fresh ground beef stored at 5°C and sampled daily until spoilage occurred. Results of this study showed that when the Eh values dropped to −60 mV, the total bacterial counts at 5 and 30°C were 10^7 per g. Two days after this count level was reached, spoilage was evidenced by organoleptic tests and methylene blue reduction (MBR). A second market survey on ground beef from retail stores was conducted in 1975. Results of Eh, total plate counts, and MBR tests closely paralleled the result of the survey made twenty years previously, emphasizing that low Eh values can be correlated with high bacterial counts and short MBR times. Coupled with two market surveys conducted twenty years apart, the controlled study by two persons using different equipment, but resulting in almost identical results, points up the strong possibility of the use of Eh as a rapid, accurate field method of determining total bacterial count in ground meat.

INTRODUCTION

RESEARCH to find a field method to indicate the quality of ground beef faster than total bacterial counts has been going on for more than sixty years. In this span of time, several

*The authors wish to express their sincere appreciation to Dr. R.R. Hocking of the Statistical Services Department of Mississippi State University for the statistical analyses.

methods have been proposed (Table 7-I) which claim to be indicators of spoilage. While some methods compare favorably with organoleptic criteria of spoilage, reviews by Turner (1960), Pear-

TABLE 7-I.

METHODS PROPOSED FOR INDICATING SPOILAGE OF GROUND BEEF

Bacteriological Methods

a. Total bacterial counts.	Ayres (1955); Bates and Highlands (1934); Brewer (1925); Duitschaever et al. (1973); Elford (1936); Fitzgerald (1947); Foltz (1941); Geer et al. (1933); Hobbs (1959); Kirsch et al. (1952); Le Fevre (1917); Nickerson et al. (1959); Rogers and McClesky (1957, 1961); Saffle et al. (1961); Weinzirl and Newton (1915).
b. Analysis for specific groups.	Ayres (1960a,b); Duitschaever et al. (1973); Gardner (1965); Halleck et al. (1958); Hoffstadt (1924); Jay (1967); Jaye et al. (1962); Kirsch et al. (1952); Thiculin et al., (1966).

Indirect bacteriological test

a. Methylene blue reduction.	Jensen (1954); Rogers et al. (1958); Rogers and McClesky (1961); Turner (1960).
b. Resazurin reduction.	Baker (1966); Proctor and Greelie (1939). Saffle et al. (1961); Turner (1960).
c. Biological oxygen demand (BOD)	Cox (1950); Turner (1960).

Tests for bacterial by-products

a. Ammonia and volatile bases	Folin and Bell (1917); Gardner and Stewart (1966); Ockerman et al. (1969); Official Methods of Analysis of the Association of Official Agricultural Chemists (1965); Pearson (1967, 1968a); Turner (1960); Wintors and Wintors (1949).
b. Free amino acids	Gardner and Stewart (1966); Jay and Kontou (1967); Turner (1960).
c. d-amino acids	Saffle et al. (1961); Turner (1960).
d. Free fatty acids	Pearson (1967, 1968d).
e. Indole	Official Methods of Analysis of the Association of Official Agricultural Chemists (1965); Turner (1960).
f. H_2S	Turner (1960); Wintors and Wintors (1949).

TABLE 7-I (Cont.)

Physical tests	
a. Acidimetric (pH or titration)	Barnes and Ingram (1955); Broumand et al. (1958); Rogers and McClesky (1961); Shelef and Jay (1970a,b); Turner (1960).
b. Meat swelling	Shelef and Jay (1969); Wierbicki et al. (1962, 1963).
c. Viscosity	Shelef and Jay (1969); Turner (1960).
d. Extract release volume	Jay (1964a,b); Kontou et al. (1966); Lowis (1971); Pearson (1968b).
e. Water-holding capacity	Hamm (1960); Jay (1965).
Organoleptic test	
a. Color	Brody (1970); Kontou et al. (1966); Lowis (1971).
b. Taste	Kontou et al. (1966); Lowis (1971).
c. Smell	Ayres (1960a,b); Kirsch et al. (1952); Kontou et al. (1966); Lowis (1971); Pearson (1967, 1968a,b); Saffle et al. (1961).
d. Slime	Ayres (1960a,b); Clauss et al. (1957); Kontou et al. (1966); Lowis (1971).

son (1968a), and Ingram and Dainty (1971) reveal that most tests indicate only advanced stages of spoilage.

Twenty years ago, the senior author made preliminary tests on the hypothesis that a direct measurement of the oxidation-reduction potentials of ground meat would correlate with the bacterial population. The results were promising, and a part of these data was published in abstract form (Rogers et al., 1958). While there has been reference to the correlation of Eh to total counts in horse meat (Barnes and Ingram, 1955), no other work has been reported applying this as a method of obtaining a rapid estimate of the bacterial counts in ground beef. This probably is the result of the assumption that Eh determinations would be difficult to duplicate in a complex system such as ground meat (Pearson, 1968a).

The objectives of this investigation were to (a) follow the changes in bacterial concentration, Eh, methylene blue reduction time, and organoleptic parameters during storage of freshly ground beef and (b) evaluate the merits of Eh as a rapid field method of estimating bacterial content of ground beef samples obtained from retail markets.

MATERIALS AND METHODS

Samples

Studies Conducted under Controlled Conditions

The Animal Science Department of Mississippi State University furnished ground beef samples from a steer which was slaughtered and chilled for two days under normal conditions. The meat was processed and wrapped in conventional fashion. Forty-five 1 lb. packages were numbered sequentially and stored at 5°C in a walk-in cooler. Three random samples were removed and analyzed daily for thirteen days.

Market Survey

Two separate market surveys were conducted. In 1955, twenty-four samples were collected from retail stores in Baton Rouge, Louisiana. In 1975, thirty-six samples were collected from retail markets in Starkville, Mississippi; all samples were at least 1 lb. in size and the analyses were begun within one hour of collection.

Analyses

Eh Determinations

In the 1955 study, Eh determinations were made with a Beckman pH meter using a platinum electrode and a reference electrode. In the 1975 study, a Corning Model 10® pH meter with an Orion Model 96-78® combination platinum electrode and a Beckman Century SS-1® pH meter with a Beckman Model 39516® combination electrode were used for all Eh determinations. The combination electrodes consisted of a platinum disc inlaid in glass, and a Ag/AgCl reference electrode; the latter is surrounded by an electrolyte solution of 4M KCl saturated with AgCl. Standardization of electrodes was as described in the 1972 Annual Book of ASTM Standards.

Tests were conducted by removing the wrapper from the ground beef and inserting the electrodes to the center of the sample, a depth of approximately 3.5 cm. Since Eh is expressed as millivolts on the pH meter and the potential is the emf difference between the constant voltage reference electrode and the platinum electrode when both are immersed in the test solution, Eh was recorded in millivolts after the needle of the pH meter became stable (1.5 to 2.0 min.). Three readings were obtained from each sample using each of the two pH meters; the six readings were averaged. Between samples, electrodes were cleaned in accordance with the 1972 Annual Book of ASTM Standards.

Total Plate Counts and Methylene Blue Reduction Test (MBR)

After Eh determinations were completed, the surface of the meat was scraped away with a sterile spoon. A 50 g sample was removed from the center of the ground meat and blended in 450 g of sterile physiological saline for three minutes at high speed in a Waring Blender.® Dilutions for the total plate counts and the MBR tests were prepared from the above 1/10 mixture.

Total plate count analyses were done using Difco Total Plate Count agar medium and the pour plate technique as described by the Manual of Microbiological Methods (1957). The 5°C plates were incubated ten days and the 30°C plates were incubated for three days.

MBR tests were conducted on 5 ml samples in 16 × 125 mm screw-capped test tubes using dilutions of 1/10, 1/20, 1/40, and 1/80. The diluent used in the last three dilutions was sterile skim milk. One ml of methylene blue solution (two tablets of standard methylene blue thiocyanate added to 200 ml boiling water and cooled) was added to each dilution and mixed by inversion of the tube. The tubes were placed in a 37°C incubator and observed every thirty minutes. A test was considered positive when methylene blue reduction was evident in two thirds of the tube.

Organoleptic Test

Color, slime, smell, and taste were evaluated by a panel of four persons from the Animal Science Department and the Mississippi State Chemical Laboratory, Mississippi State, Mississippi. Results were recorded as "+" for acceptable and "−" for unacceptable.

Statistical Analyses

These analyses were done on the linear or quadratic Least Squares Program by the Statistical Services Department, Mississippi State University, under the direction of Dr. R. R. Hocking.

RESULTS AND DISCUSSION

Studies Conducted under Controlled Conditions

The first series of tests was directed toward comparing the changes in the oxidation-reduction potential of ground beef during storage to the total bacterial plate counts, methylene blue reduction times, and organoleptic parameters (color, smell, slime, and taste). Therefore, samples of freshly processed ground beef, stored under rigidly controlled conditions, were examined daily. As indicated in Table 7-II, the Eh decreased with increased length

TABLE 7-11

RESULTS FROM GROUND BEEF SAMPLES STORED UNDER CONTROLLED CONDITIONS AT 5°C.

Days Storage	Eh (mV)	MBR* (h)	Color**	Smell**	Slime**	Taste**	Total Plate Count 30°C (No./g)	5°C (No./g)
0	28	>4.0	+	+	+	+	2.2×10^5	2.5×10^5
1	56	>4.0	+	+	+	+	4.6×10^4	6.6×10^4
2	53	>4.0	+	+	+	+	2.9×10^5	2.5×10^5
3	35	>4.0	+	+	+	+	3.8×10^5	2.1×10^5
4	26	>4.0	+	+	+	+	1.0×10^5	1.2×10^5
5	18	>4.0	+	+	+	+	3.6×10^5	3.5×10^5
6	22	>4.0	+	+	+	+	1.4×10^6	1.7×10^6
7	—10	>4.0	—	+	+	+	2.6×10^6	3.1×10^6
8	—60	>4.0	—	+	+	+	1.9×10^7	1.9×10^7
9	—90	4.0	—	+	+	+	3.3×10^7	4.0×10^7
10	—190	2.5	—	+	+	+	1.0×10^8	1.0×10^8
11	—120	2.5	—	—	—	—	1.2×10^8	1.2×10^8
12	—220	2.0	—	—	—	—	2.4×10^8	1.9×10^8
13	—250	2.0	—	—	—	—	1.2×10^8	1.2×10^8

Bacterial counts and Eh values are averages obtained from triplicate samples.

* MBR = Methylene Blue Reducation Time;
Numbers = hours required for 66 percent of tube to show reduction

** + = acceptable
— = unacceptable

of storage with a concurrent increase in total bacterial plate count at both 30°C and 5°C. Rogers and McClesky (1961) suggested that a methylene blue reduction time of four hours indicated poor keeping quality of ground beef; in the current tests, this situation occurred after nine days of storage. From an organoleptic stand-point, color, which is considered to be the least affected of all physical parameters by bacterial action, remained acceptable through six days of storage and the other parameters (smell, slime, and taste) were still acceptable up to ten days of storage. It is noteworthy to point out that the bacterial populations had in-creased to significant numbers just prior to detection by organo-leptic tests and methylene blue reduction time.

Figure 7-1 illustrates the changes in bacterial population and

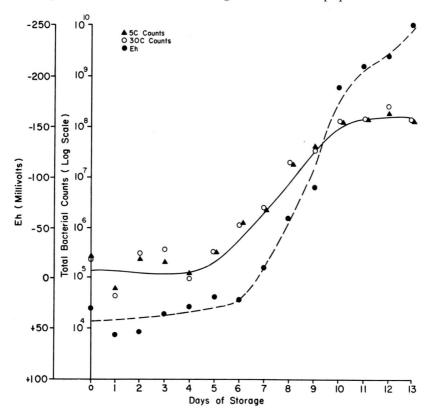

Figure 7-1. Eh and total plate counts (5 and 30°C) on samples stored under controlled conditions for thirteen days.

Eh with length of storage. The shape of the curve for the bacterial counts is typical for a bacterial growth curve and is paralleled rather closely by the changes in oxidation-reduction potential up to approximately ten days. As indicated in Table 7-II, the meat was unacceptable after this period of time from an esthetic standpoint; nevertheless, the oxidation-reduction potential continued to decrease while the bacterial count remained essentially constant. This indicates that the oxidation-reduction potential is continuing to reflect the cumulative effect of the bacterial action even though the total number of organisms involved had stabilized.

The relationship between bacterial counts and Eh is indicated in Figure 7-2. There appears to be a direct correlation between numbers of organisms and Eh of the meat up to an Eh of approximately -100 mV. Thereafter, Eh continued to decrease while the bacterial count remained essentially constant. In this context, the maximum limit of 10^7 bacteria per g permitted by the State of Oregon food regulations is indicated by an Eh reading of -60 mV.

Market Survey

The above data strongly support the hypothesis that Eh determinations are well suited to estimating the total bacterial content in ground beef, and the question remains as to the utility of the method as a practical field procedure. To help answer this question, the results of two separate market surveys are described below: one in Baton Rouge, Louisiana in 1955, and the other in Starkville, Mississippi in 1975. In both surveys it was found that significant bacterial populations were present in the meat before a methylene blue reduction time of four hours or less was observed and there was essentially no correlation of MBR time to bacterial count.

The statistical analyses for the two studies consisted of comparing bacterial counts to the Eh readings. Preliminary analysis indicated that the logarithm of bacterial count was the appropriate dependent variable and this was borne out by the residual analysis on the final results. Linear and quadratic models were considered in all cases. The choice between the two was based on

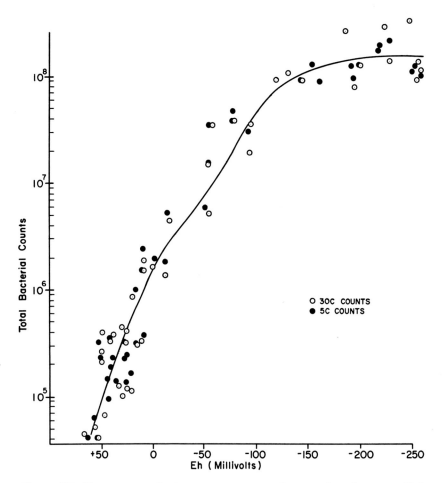

Figure 7-2. Eh versus total plate counts on samples stored under controlled conditions for thirteen days.

the residual analyses followed by a test of significance of the quadratic coefficient at the 5 percent level. Analysis of normal plots of residuals justified the assumption of normal errors.

Figure 7-3 is a quadratic least squares plot of Eh versus the total bacterial count at 30°C, and Figure 7-4 is a plot at 5°C. It may be observed that these plots are essentially identical, thus reflecting a situation similar to that observed in the controlled

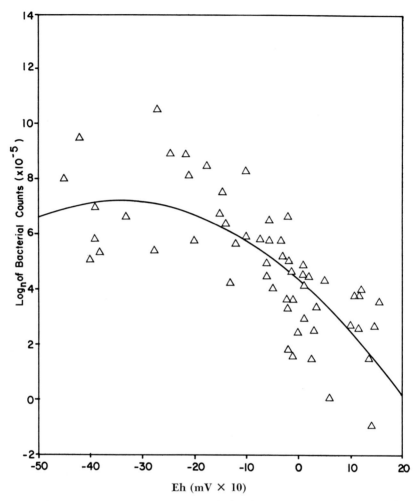

Figure 7-3. Quadratic least squares analysis of total bacterial counts (30°C) versus Eh from the 1955 and 1975 market survey studies.

study. It is obvious that the reason a quadratic least squares plot more accurately describes the data is the fact that the samples having greatly reduced Eh values do not have increasingly high bacterial counts. This was expected from the study conducted under controlled conditions where it was shown that the Eh values continue to be reduced after the bacterial counts have become

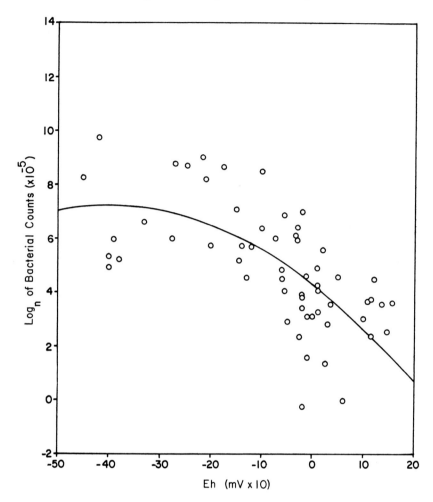

Figure 7-4. Quadratic least squares analysis of total bacterial counts ($5°C$) versus Eh from the 1955 and 1975 market survey studies.

stabilized. Therefore, samples more reduced than -300 mV were deleted, and the linear least squares plots of the data are given in Figures 7-5 and 7-6.

The plots for the 1955 survey are almost identical with the plots for the 1975 survey, which tends to add weight to the practicality of this method for routine field use. Thus, an Eh of -60

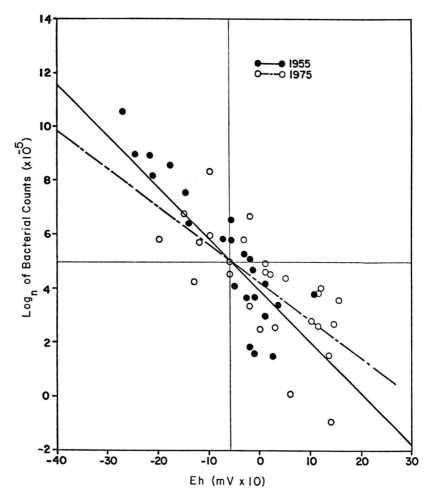

Figure 7-5. Linear least squares analyses of total bacterial counts (30°C) versus Eh. All samples more reduced than −300 mV are excluded.

mV or greater would serve as a division between ground beef with low or acceptable bacterial counts ($< 10^7$ bacteria/g) while lower Eh values would indicate unacceptably high bacterial counts ($> 10^7$ bacteria/g). Based on this standard, the bacterial content of 91 percent of the 1975 market survey samples and 89 percent of the 1955 market survey samples were accurately predicted by the

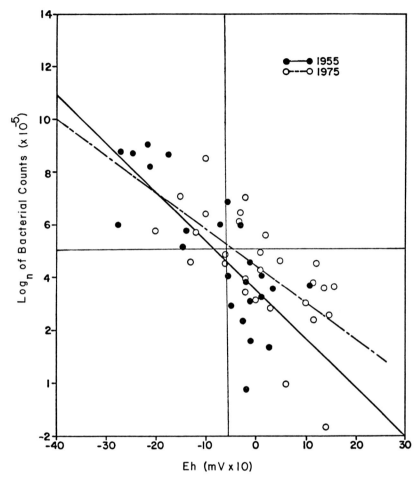

Figure 7-6. Linear least squares analyses, total bacterial counts (5°C) versus Eh. All samples more reduced than −300 mV are excluded.

Eh values. Correspondingly, bacterial counts at 5°C were accurately predicted in 91 percent of the 1975 samples and 97 percent of the 1955 samples on the basis of Eh determinations.

CONCLUSION

It has been demonstrated that decomposition of ground beef during storage, as evidenced by increases in bacterial content, can be rapidly estimated in minutes by the direct measurement of the oxidation-reduction potential of the sample.

In two market surveys conducted twenty years apart and in different geographic locations there was excellent correlation between the bacterial counts and Eh. Furthermore, the linear least square plot for the 1975 survey was essentially identical to the plot for the 1955 survey, this indicating not only the similarity of results but also the dependability of the method. Eh values of —60 mV were indicative of bacterial counts of greater than 10^7 cells/g.

While this paper is not intended to minimize the value of plate counts or other suggested methods of testing the quality of ground beef, Eh determinations can be done in the field in a matter of minutes as compared to three to ten days for plate counts or even four hours required for MBR tests. Additionally, the saving in terms of labor, supplies, and equipment, by utilization of Eh instead of conventional plate count methods, in this day of increasing costs cannot be overemphasized.

REFERENCES

1972 Annual Book of ASTM Standards. American Society for Testing and Materials, Philadelphia, p. 341.

Ayres, J.C.: 1955. Microbiological implications in the handling, slaughtering and dressing of meat animals. *Adv Food Res, 6:*109.

Ayres, J.C.: 1960a. The relationship of organisms of the genus *Pseudomonas* to the spoilage of meat, poultry, and eggs. *J Appl Bact, 23:*471.

Ayres, J.C.: 1960b. Temperature relationships and some other characteristics of the microbial flora developing on refrigerated beef. *Food Res, 24:*1.

Baker, K.R.: 1966. Adaptation of resazurin test to meat products. *Food Manuf, 41:*49.

Barnes, E.M. and M. Ingram: 1955. Changes in the oxidation-reduction potential of the sternoaphalicus muscle of the horse after death in relation to the development of bacteria. *J Sci Food Agric, 6:*448.

Bates, P.K. and M.E. Highlands: 1934. The determination of storage conditions. *Refrig Eng, 27:*299.

Brewer, C.M.: 1925. The bacteriological content of market meats. *J Bact, 10:*543.

Brody, A.L.: 1970. Shelf life of fresh meat. *Modern Package, 43:*81.

Broumand, H., C.D. Ball, and E.F. Stier: 1956. *Food Technol,* Champaign, *12:*159.

Clauss, W.E., C.O. Ball, and E.F. Stier: 1957. Factors affecting quality of prepackaged meat. I. Physical and organoleptic test. C. Organoleptic and miscellaneous physical characteristics of product. *Food Technol, 11:* 363.

Cox, H.E.: 1950. *The Chemical Analysis of Foods,* 4th ed. Wiley, New York. p. 215.

Duitschaever, C.L., D.R. Amott, and D.H. Bullock: 1973. Bacteriological quality of raw refrigerated ground beef. *J Milk Food Technol, 36:*375.

Elford, W.C.: 1936. Bacterial limitations in ground fresh meat. *Am J Public Health, 26:*1204.

Fitzgerald, G.A.: 1947. Are frozen foods a public health problem? *Am J Public Health, 37:*695 .

Folin, O. and R.D. Bell: 1917. *J Biol Chem, 29:*329.

Foltz, V.D.: 1941. A bacteriological study of ground meat. *J Bact, 42:*289.

Gardner, G.A.: 1965. The aerobic flora of stored meat with particular reference to the use of selective media. *J Appl Bact, 28:*252.

Gardner, G.A. and D.J. Stewart: 1966. Changes in the free amino and other nitrogen compounds in beef muscle. *J Sci Food Agric, 17:*491.

Geer, L.P., W.T. Murray, and E. Smith: 1933. Bacterial content of frosted hamburger steak. *Am J Public Health, 23:*673.

Halleck, F.E., C.O. Ball, and E.F. Stier: 1958. Factors affecting quality of prepackaged meat. IV. Microbiological studies. *Food Technol, 12:*197.

Hamm, R.: 1960. Biochemical of meat hydration. *Adv Food Res, 10:*355.

Hobbs, B.C.: 1959. Sampling and examination of foodstuffs and interpretation of results. *Munic Eng Motor Pub, 136:*469.

Hoffstadt, R.E.: 1924. Bacteriological examination of ground beef. 1. Relation of bacterial count and aerobic species present to spoilage. *Am J Hyg, 4:*33.

Ingram, M. and R.H. Dainty: 1971. Changes caused by microbes in spoilage of meats. *J Appl Bact, 34:*21.

Jaye, M., R.S. Kittaka, and Z.J. Ordal: 1962. The effect of temperature and packaging material on the storage life and bacterial flora of ground beef. *Food Technol, 16:*95.

Jay, J.M.: 1964a. Release of aqueous extracts by beef homogenates, and factors affecting release volume. *Food Technol,* Champaign, *18:*129.

Jay, J. M.: 1964b. Beef microbial quality determined by extract-release volume (ERV) *Food Technol,* Champaign, *18:*133.

Jay, J.M.: 1965. Relationship between waterholding capacity of meats and microbial quality. *Appl Microbiol, 13:*120.

Jay, J.M.: 1967. Nature, characteristics, and proteolytic properties of beef spoilage bacteria at low and high temperatures. *Appl Microbiol, 15:*943.

Jay, J.M. and K.S. Kontou: 1967. Fate of free amino acids and nucleotides in spoiling beef. *Appl Microbiol, 15:*759.

Jensen, L.B.: 1954. *Microbiology of Meats,* 3rd ed. U. Garrard Press, Champaign.

Kirsch, R.H., F.F. Berry, C.L. Baldwin, and E.M. Foster: 1972. The bacteriology of refrigerated ground beef. *Food Res, 17:*495.

Kontou, K.S., M.C. Huyck, and J.M. Jay: 1966. Relation between sensory test scores, bacterial numbers and ERV on paired raw and cooked ground beef from freshness to spoilage. *Food Technol,* Champaign, *20:*128.

LeFevre, E.: 1917. A bacteriological study of hamburger steak. *Am Food J, 12:*140.

Lowis, M.J.: 1971. The role of extract release volume in a rapid method for assessing the microbial quality of pork and beef. *Food Technol, 6:* 415.

Manual of Microbiological Methods. 1957. Society of American Bacteriologists. McGraw, New York.

Nickerson, J.T.R., B.E. Proctor, and E.J. Robertson: 1956. Microbiology of frozen foods. *Air Conditioning and Refrigerating Data Book,* vol. 1, p. 21. The Am. Soc. Refrig. Engineers, New York.

Ockerman, H.W., V.R. Cahill, H.H. Weiser, C.E. Davis, and J.R. Siefker: 1969. Comparison of sterile and inoculated beef tissue. *J Food Sci, 34:*93.

AOAC: 1965. *Official Methods of Analysis of the Association of Official Agricultural Chemists.* 10th ed. AOAC, Washington, D.C.

Pearson, D.: 1967. Assessing beef acceptability. *Food Manuf, 42:*42.

Pearson, D.: 1968a. Assessment of meat freshness in quality control employing chemical techniques. A review. *J Sci Food Agric, 19:*357.

Pearson, D.: 1968b. The correlation of the extract-release volume of stored beef with other spoilage values. *Food Technol, 3:*207.

Pearson, D.: 1968c. Application of chemical methods for the assessment of beef quality. I. General consideration, sampling and the determination of basic components. *J Sci Food Agric, 19:*364.

Pearson, D.: 1968d. Application of chemical methods for the assessment of beef quality. III. Methods related to fat spoilage. *J Sci Food Agric, 19:*553.

Proctor, B.E. and B.G. Greenlie: 1939. Redox potential indicators in quality control of foods. *Food Res, 4:*441.

Rogers, R.E. and C.S. McCleskey: 1957. The bacteriological quality of ground beef in retail markets. *Food Technol, 11:*318.

Rogers, R.E., L.R. Brown, and C.S. McCleskey: 1958. Criteria of freshness in ground beef. *Bact Proc, A63:*25.

Rogers, R.E. and C.S. McCleskey: 1961. Objective tests for quality of ground beef. *Food Technol,* Champaign, *15:*210.

Saffle, R.L., K.N. May, H.A. Mamid, and J.D. Irby: 1961. Comparing three rapid methods of detecting spoilage in meat. *Food Technol, 15:*465.

Shelef, L.A. and J.M. Jay: 1969. Relationship between meat swelling, viscosity, extract release volume, and water-holding capacity in evaluating beef microbial quality. *J Food Sci, 34*:532.

Shelef, L.A. and J.M. Jay: 1970a. Use of a titrimetric method to assess the bacterial spoilage of fresh meats. *Bact Proc,* 6.

Shelef, L.A. and J.M. Jay: 1970b. Use of a titrimetric method to assess the bacterial spoilage of fresh beef. *Appl Microbiol, 19:* 902.

Thiculin, G., J. Pantaleon, and R. Rosset: 1966. Contributions à l'étude des germes airobies psychrotrophes des viandes hachies. *Inst Pasteur* (Lille), *17:*131.

Turner, A.: 1960. Assessing meat spoilage in the laboratory. *Food Manuf, 35:*386.

Weinzirl, J. and E.B. Newton: 1915. The fate of bacteria in frozen meat held in cold storage and its bearing on a bacteriological standard for condemnation. *Am J Public Health, 5:*833.

Wells, F.E.: 1956. Resazurin reduction tests for shelf life estimations of poultry meats. *Food Technol,* Champaign, *13:*584.

Wierbicki, E., M.G. Tiede, and R.G. Burrell: 1962. Determination of meat swelling as a method for investigating the water-binding capacity of muscle proteins with low water-holding forces. *Fleischwirtschaft, 14:* 951.

Wierbicki, E., M.G. Tiede, and R.G. Burrell: 1963. Determination of meat swelling as a method for investigating the water-binding capacity of muscle proteins with low water-holding forces. II. Application of the swelling methodology. *Fleischwirtschaft, 15:*404.

Wintors, A.L., and K.B. Wintors: 1949. *The Structure and Composition of Foods.* Wiley, New York.

Chapter 8

AUTOMATED FLUORESCENT ANTIBODY TEST FOR *SALMONELLA**

T. E. MUNSON, J. P. SCHRADE, AND N. B. BISCIELLO, JR.

Abstract

A prototype automated system employing direct fluorescent antibody (FA) was evaluated for ability to screen foods and feeds for *Salmonella.* Samples were subjected to the usual enrichment steps in selenite cystine and tetrathionate broths and then transferred into fresh selenite cystine for a 4 h "post-enrichment" to dilute out possible background fluorescence due to food debris. The system operates as follows: Samples of post-enrichment broth are placed in individual holders in a processor where they are applied automatically to filters on plastic slides. The slides pass to stations where cells are washed to remove broth, stained with polyvalent FA conjugate, incubated for 15 min at 37°C, washed to remove unreacted antibody, and air dried. The slides are stacked in a cassette and placed in an automatic reader where cells are irradiated with an excitation light source. The reader, using an arbitrary scale of 0 to 200 automatically measures fluorescence and records it for each slide on a digital display and on a paper tape print out.

FA results were compared with those obtained by the conventional methods of The Association of Official Analytical Chemists (AOAC). Initially, 167 samples of milk powder, dried yeast, and imported frog legs were examined. Good correlation between the two methods was found with all samples but frog legs. Problems with this product were attributed to competing microorganisms in the post-enrichment broth and to equipment design. These problems were corrected and with an additional 188 samples, excellent correlation was obtained between the AOCA and automated methods. When a value of 80 on the fluorescence scale was used as the cut-off between FA positive and negative samples there were no false FA nega-

*The authors would like to thank Dr. James D. Macmillan, Chairman of the Department of Biochemistry and Microbiology, Rutgers University, New Brunswick, New Jersey for his invaluable assistance in the preparation of the manuscript.

tive samples and 5 percent false FA positive samples. Thus, all FA positives must be confirmed biochemically and serologically.

Introduction

THE direct fluorescent antibody (FA) technique is a specialized analytical procedure whereby antigen-antibody reaction can be readily detected by fluorescent microscopy. The technique is based upon the finding that fluorescent dyes such as fluorescein isothiocyanate can be linked to antibodies without destroying their specific immunological reactivity. The procedure is particularly useful for rapid identification of bacteria in mixed populations or in infected materials. Smears on a slide are covered with a solution of fluorescein-labeled antibody specific for the suspected bacteria. After rinsing to remove unbound antibody the smear is examined by fluorescent microscopy, and the suspected organisms are visualized by the intense fluorescence of the attached dye.

Publications by Coons et al. (1941, 1942) represent the first application of FA techniques in microbiology. These papers described the preparation of fluorescein protein conjugates and their use in detecting pneumococcal antigen in tissues of infected mice. FA test systems are now available for the clinical detection of *Streptococcus, Salmonella, Neisseria gonorrhoeae, Treponema pallidum, Toxoplasma gondii,* and a large number of other pathogenic organisms. Various adaptations of the FA procedure have been proposed for detection of *Salmonella* in products such as poultry, eggs, raw meat, etc., by Fantasia et al. (1969, 1975); Georgala et al. (1964, 1965) ; Haglund et al. (1964) ; Insulata et al. (1967) ; Schulte et al. (1968) ; and Silliker et al. (1966) .

In 1975, Fantasia et al. of the United States Food and Drug Administration evaluated the FA technique and found that polyvalent FA preparations presently available commercially in the United States are quite suitable for screening foods and feeds for *Salmonella.* Although these preparations can cross react with some *Citrobacter sp.* and *Escherichia coli,* in actual analyses of over 4,000 food and feed samples a total of 666 samples were positive by the FA method as compared to 619 by the cultural method of the Association of Official Analytical Chemists (AOAC, 1975). This represents a false positive rate of about 7 percent due to cross reactions with non-salmonellae. Only four samples positive by the AOAC method were negative by the FA method. As a part of the evaluation by Fantasia et al. (1975), the FA method was subjected to collaborative study by twenty-two other governmental and industrial laboratories. The data obtained supported use of the FA method and as a result it was approved as official first action by AOAC for use in the routine examination of samples suspected of *Salmonella* contamination. It was stipulated, because of the false positive rate, that the method should only be used to screen for *Salmonella* and that all FA positive samples should be confirmed by AOAC culture, biochemical and serological procedure. The collaborative study was designed to compare results from eleven analysts experienced with FA methodology with those having little or no experience in this area. In a total of 200 samples, the experienced group found 127 positives, of which 125 were confirmed as positive by the cultural AOAC methods. Thus, there were no false negatives and only two false positives. The inexperienced group, however, reported nine false negatives and two false positives. These results showed clearly that experience and training are required in order to run and interpret the FA results by conventional manual procedures. The FA method is a qualitative procedure and microscopic examinations must be evaluated somewhat subjectively.

Until 1975, very little work had been reported on ways to reduce the subjectivity of the FA method through automation. The only equipment described was the microscope photometers used to measure the degree of fluorescence. These photometers usually are capable of reading only one microscopic field at a time and the reading of additional fields can require as much time as thorough manual examinations. Sensitivity is poor due to limitations in the microscope optics. In the early 1970s, Organon Diagnostic (formerly Aerojet Medical and Biological Systems) developed automated equipment to prepare and read FA slides.

This equipment does not depend on microscope optics for measuring the fluorescence of the stained specimen. A prototype model of the automated system was evaluated by our laboratory (Munson et al., 1976). An operational procedure was developed for automated analysis of *Salmonella* in foods and feed and the results were compared to those obtained by conventional AOAC methods. This equipment was also evaluated by Thomason et al. (1975) at the Center for Disease Control, whose results indicated that the automated FA correlated well with the manual FA method for processed food but not for raw meat products and environmental samples.

Materials and Methods

AUTOMATED EQUIPMENT. The Automated Bioassay System® has two major components, a slide processor and a slide reader (Figure 8-1). The slide processor automatically adds 0.1 ml of culture broth onto a plastic slide (Figure 8-2) having a centrally located grid which supports a 0.8 μ membrane filter disc. The disc is held by white plastic tape and the filtering area is 4 mm in diameter. Automatically, each slide is transported to a station where suction is applied to the lower surface of the slide for 20 sec. During this time a wash solution, consisting of phosphate buffered saline, pH 7.2 (FA buffer, Difco) with 0.5 ml/1 of Tween 80®, is drawn under vacuum through the filter to wash the cells. Approximately 0.1 ml of titred fluorescent antibody is automatically added to the filter. The slides then pass through a humidified, temperature controlled chamber where they incubate at 35°C for about 15 min. They are washed twice, to remove unreacted fluorescent antibody. The slides are dried by hot air and stacked automatically in a second cassette. The first slide takes 21 min to process, followed by the next slides at 30 sec intervals, i.e. 120 slides per hour (h).

The reader is a fluorimeter having excitation and emission filters for the analysis of fluorescein isothiocyanate. A cassette of processed slides is placed in the reader. Each slide is automatically transported to a station where the filter is moistened with wash solution and then to a reading station where it is irradiated with

Figure 8-1. Prototype model of the Automated Bioassay System manufactured by Organon Diagnostics.   Left: Slide   Processor.   Right: Fluorescence Reader.

Figure 8-2. Plastic filter slide designed for use with the Automated Bioassay System.

an excitation light source. The degree of fluorescence is read on an arbitrary scale of 0 to 200. Readings are displayed on a display panel and printed on a paper tape. One slide is read and recorded every 10 sec (360 per h) and then stacked into a second cassette.

FLUORESCENT ANTIBODY. Difco polyvalent fluorescent antibody (Bacto-FA Salmonella Poly) was used at a working titre of 1:8 for both the manual and automated procedures. For the automated system, Tween 80 (2% v/v) was added to the fluorescent antibody to prevent absorption onto the plastic filter slides. Prior to use, the fluorescent antibody preparation was filtered to remove particulate matter. After filtration the antibody must not come in contact with metal or it will precipitate in the equipment and cause false positive reactions.

SAMPLES. Samples were collected from commercial sources as routine surveillance samples by the FDA, New York District. Some samples had been tested previously by AOAC cultural methods and were stored at room temperature until analyzed with the automated system; all other samples were tested with the automated system immediately upon receipt.

CULTURES, MEDIA, AND ANTISERA. *Salmonella typhimurium* (ATCC 14028) was used for the positive controls with the automated system. Cultures were maintained in FAS broth (Difco) and transferred daily to provide a ready source of actively growing cells. Enrichment, plating, and all other media were obtained from both Difco and BBL and were used interchangeably regardless of the source. The antisera used to determine the *Salmonella* group for isolates were obtained from Difco.

ANALYSIS BY AOAC SALMONELLA METHODS. Samples examined by the automated system were also analysed by culture, biochemical and serological procedures as described in the *AOAC Compendium* (AOAC, 1975). The samples were pre-enriched according to the *FDA Bacteriological Analytical Manual* (Olson, 1972). The methods call for a pre-enrichment step in either lactose broth, rehydrated milk, or sterile water, depending upon product type, followed by enrichment in both selenite cystine and tetrathionate broths. After 18 to 24 h at 35°, the enrichment broths were streaked onto selective agars. Suspected colonies were picked and confirmed as *Salmonella* by biochemical and serological reactions as outlined in the methods (AOAC, 1975).

All samples were also analyzed by the manual FA method as described in the supplement to the AOAC Compendium (AOAC, 1974). The method calls for a 4 h post-enrichment step which eliminates product carryover that occasionally causes background fluorescence and also produces young cells which are very antigenic.

AUTOMATED ANALYSIS OF FOOD SAMPLES. An outline of the overall procedure which was developed for evaluating the automated system is shown in Figure 8-3. Samples of various foods were pre-enriched as previously described. The analysis by conventional AOAC methods is shown on the right hand portion of Figure 8-3.

The procedure on the left of Figure 8-3 was used to prepare samples for analysis of *Salmonella* by the automated system. A transfer was made to a second tube of selenite cystine broth which was incubated for 4 h at 35°C as a post-enrichment step. The post-enrichment cultures were streaked onto Hektoen enteric agar plates (Bisciello and Schrade, 1975) as a control on viability and then the cultures were treated with formalin to prevent contamination of the equipment. Two 0.1 ml aliquots of these broths were analyzed by the automated system, and smears on conventional microscopic slides were prepared for analysis by the manual FA method.

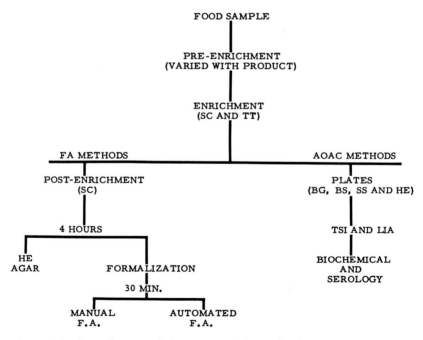

Figure 8-3. Flow diagram of the tests used in evaluation of the automated FA system. Abbreviations: SC, Selenite Cystine Broth; TT, Tetrathionate Broth; B.G. Brilliant Green Agar; BS, Bismuth Sulfite Agar; SS, Salmonella-Shigella Agar; HE, Hektoen Enteric Agar; TSI, Triple Sugar Iron Agar slants; and LIA, Lysine Iron Agar slants.

Positive and negative food samples were used as controls for each different type of product tested. The positive control was a sample of food inoculated with a 24 h broth culture of the control organisms and processed through all the enrichment steps along with the test samples. The negative food control consisted of the uninoculated product appropriately diluted in the various media without incubation between each transfer. These controls indicated whether the product was inhibitory for *Salmonella* and if product carryover caused false FA readings.

RESULTS AND DISCUSSION
Development of the Evaluation Procedure

Preliminary studies were performed to determine the fluorescence readings resulting from analysis of various dilutions of *Salmonella* cells with the automated equipment. Cell counts in the dilutions were determined by the direct microscopic method. The experiment was repeated on different days and the results are shown in Figure 8-4. The fluorescence readings did not change

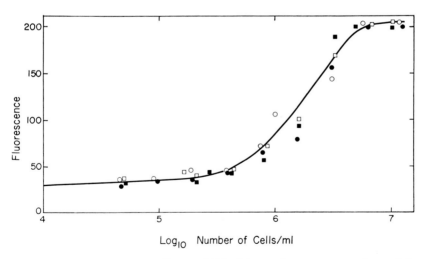

Figure 8-4. Fluorescence readings exhibited by various concentrations of *S. typhimurium* cells processed through the automated system. Each type of symbol represents results from a separate series of dilutions. Each dilution series was prepared and processed on a different day.

significantly until cell numbers approached about 8×10^5 per ml. Fair agreement in reading was obtained with the successive runs. According to the manufacturer of the equipment, about 10^6 cells are required to produce a fluorescence reading of 75. This value was confirmed in the preceding experiment.

The next step in the evaluation was to develop a cultural procedure for preparing samples for the automated system that would yield at least 10^6 *Salmonella* cells per ml and produce readings clearly in the useful range (about 50 on the scale). Various enrichment media, post-enrichment media, and incubation times were evaluated. Of the media tested, both selenite cystine and tetrathionate broth proved effective as enrichment broths (other media tested were tryptic soy tryptose broth and FA Salmonella broth). Tetrathionate broth did not prove to be as effective as selenite cystine broth for use in a post-enrichment step. Highest fluorescence readings were obtained when an 18 to 24 h enrichment broth culture was transferred to fresh selenite cystine broth and incubated for 4 h at 35°C.

Analysis of Food Samples

Early in this work it was found that the automated system performed well with samples of dry milk powder and yeast powder. In comparison to other kinds of foods contaminated with *Salmonella*, these two products usually contain fewer numbers of non-*Salmonella* organisms. An example of typical results obtained with samples of dry milk is shown in Figure 8-5. The samples were confirmed positive or negative by the AOAC cultural methods. The lowest fluorescence reading obtained for the confirmed positive samples with the automated FA analysis was about 20 units higher than the highest reading for the confirmed negative samples.

Imported frog legs, which contained a high level of *Salmonella* and also many competitive non-*Salmonella* organisms, proved to be a problem. In this instance, the fluorescent readings showed no correlation with the confirmed positive and negative samples (Figure 8-6). The fluorescence readings for confirmed positive samples ranged throughout the entire scale. All confirmed posi-

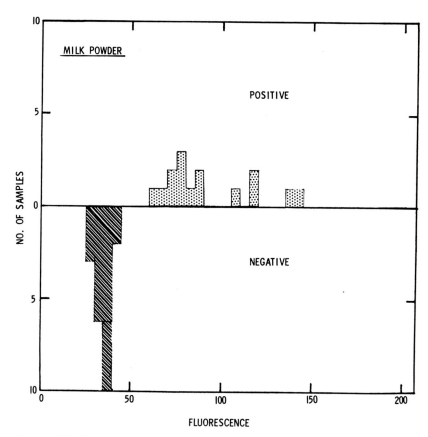

Figure 8-5. Fluorescence reading obtained by analysis of thirty-six samples of dry milk powder for *Salmonella* by the automated procedure. Duplicate slides were prepared for each post-enrichment broth, and the fluorescence readings were averaged. For comparison, the number of samples confirmed as positive for salmonellae by the AOAC procedure is shown above the center line, and the number of confirmed negative samples is shown below the line.

tive samples but no confirmed negatives, were positive by manual FA methods. In some instances, the competitive bacteria prevented the *Salmonella* from reaching sufficient levels to produce fluorescence readings that were within the useful area above 50.

During the analysis of frog legs a problem with the automated sample delivery system was noticed. The volume of post-enrich-

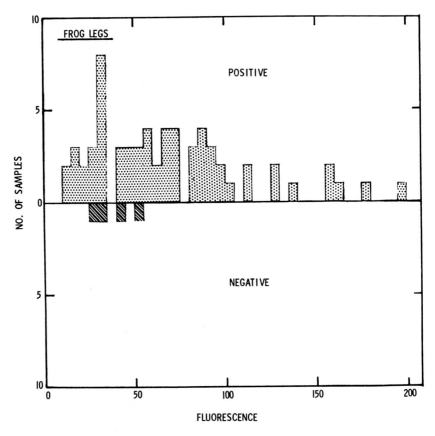

Figure 8-6. Results for the automated FA analysis of sixty-eight samples of imported frog legs before modification of procedures and equipment. Comparison as in Figure 8-5.

ment broth applied to the slide was sometimes lower than 0.1 ml. In addition, as the slides passed through the equipment, they occasionally became flooded at the wash station. These problems caused substantial variation between readings obtained for duplicate samples of the same post-enrichment culture.

Part of the problem with uneven sample delivery was caused by incomplete draining of the sample holder due to adherence of liquid to the holder walls. This was corrected by adding 0.05 ml of wash solution (which contains Tween 80) to each 0.1 ml of culture broth held in the sample wells. A mechanical problem, also connected with uneven sample delivery, was corrected by modification of the delivery system. In addition, the manufacturer modified the reader to provide increased sensitivity so that fewer cells were required to obtain a reading above 50. However, a solution to the occasional flooding problem had not been found at this stage.

After the above corrections were made, an additional 116 samples of frog legs were analyzed. The readings (Figure 8-7) for confirmed positive samples were now separated from the read-

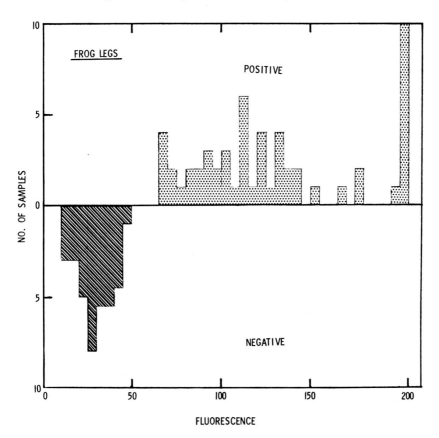

Figure 8-7. Results of the automated FA analysis of 116 samples of frog legs after modification of procedures and equipment. Comparison as in Figure 8-5.

ings for confirmed negative samples and there was less variation between duplicate samples than in the earlier frog leg study. Most duplicate samples varied no more than 10 units. If the variation was greater than 25, a microscopic examination of the slide was performed and usually showed fluorescent debris or precipitated FA conjugate.

For the final stage of this evaluation study, a solution was found to prevent flooding of the filter slides. The problem was traced to an improper design of the filter support grid, in the plastic slide, which reduced the effective filtering area by one fourth. This was corrected when the manufacturer supplied new slides with an improved grid.

The automated system was then tested with a larger variety of products in an effort to determine whether the equipment could be used routinely for *Salmonella* surveillance. A total of 188 samples of 10 different commodities were analysed. No attempt was made to include all of the products that are susceptible to *Salmonella* contamination. The overall results obtained with these products are shown in Figure 8-8. Of the 188 samples, nineteen were confirmed positive by AOAC tests. The majority of the

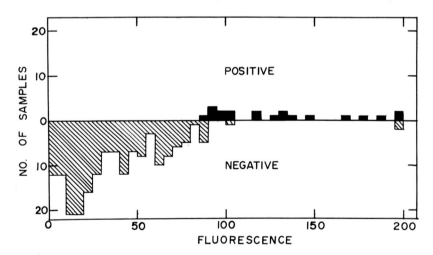

Figure 8-8. Results of the automated FA analysis of 188 samples of various commodities. Comparison as in Figure 8-5.

automated FA negative samples were well separated from the confirmed positive samples. However, there was some crossover in the range of about 8 to 100 and there were two negative samples that gave readings of 200. The objective of this study was to see if there was some number on the arbitrary fluorescence scale which would make a logical cut-off point between positive and negative samples.

By inspection of the data in Figure 8-8 it can be seen that all confirmed *Salmonella* positive samples gave a fluorescence reading of 85 or higher. A value of 80 was selected as the cut-off point to provide an additional safety factor. If all FA samples above 80 are considered FA positive and below 80 FA negative, the results in Figure 8-8 can be tabulated by commodity type as shown in Table 8-I. Of a total of 188 samples, twenty-seven would be FA positive as compared to nineteen confirmed positives by the AOAC cultural method. Thus, nine samples, or about 5 percent of the 188 tested, were false positives. This false positive rate compares well with the results of Fantasia et al. (1969) of the authors' laboratory, who analysed over 4,000 samples by the manual FA method and found that the false positive rate was about 7 percent, presumably due to cross reaction of the FA conjugate with *Citrobacter* and

TABLE 8-I

COMPARISON OF THE DATA SHOWN IN FIG. 8-8 WHEN 80 IS USED AS
THE CUT-OFF BETWEEN AUTO FA POSITIVE AND NEGATIVE
SAMPLES WITH THE AOAC CULTURAL RESULTS.

Commodity	No. of Samples	Positive Samples		Negative Samples		False Positive	False Negative
		AOAC	Auto FA	AOAC	Auto FA		
Frog Legs	34	3	6	31	28	3	
Chocolate	33	1	2	32	31	1	
Spices	29	6	8	23	21	2	
Egg Products	5	2	2	3	3		
Feeds	26	4	4	22	22		
Gums	6	1	1	5	5		
Sea Foods	12	2	5	10	8	3	
Gelatin	7			7	7		
Noodles	9			9	9		
Coconut	27			27	27		
	188	19	28	169	160	9	

some strains of *E. coli*. However, reexamination of eight samples originally designated as FA positive by the manual method and negative by the AOAC method showed that salmonellae were actually present and had been missed by the AOAC method. Thus, there is a possibility that the false positive rate is actually lower than has been stated. In any event, a false positive rate of about 5 percent, or even higher if additional safety proves necessary, does not represent a marked disadvantage to the automated method.

On the basis of these studies, the authors have concluded that the automated system provides a rapid, economical method for screening foods and feeds. On routine testing, the vast majority of samples analysed would be negative and these could be eliminated immediately from additional testing at the enrichment stage and certified *Salmonella* free. Thus, FA negative samples would not require plating on selective agars and additional biochemical and serological testing. Because of the false positive rate, all FA positives would require confirmation by additional biochemical and serological tests.

REFERENCES

AOAC: 1974. Changes in official methods of analysis made at the 88th annual meeting, October 14-17, 1974, 58 Sections 46.A03-46.A06, 417 *J Assoc Anal Chem.*

AOAC (W. Horwitz, ed.): 1975. *Official Methods of Analysis of the Association of Official Analytical Chemists,* 12th ed. Sections 46.013-46.026, pp. 906. Association of Official Analytical Chemists, Washington, D.C.

Bisciello, N.B., Jr. and J.P. Schrade: 1975. Evaluation of Hektoen enteric agar for the detection of *Salmonella* in foods and feeds. *J Assoc Off Anal Chem, 57*:992.

Coons, A.H., H.J. Creech, and R.N. Jones: 1941. Immunological properties of an antibody containing a fluorescent group. *Proc Soc Exp Biol Med, 47*:200.

Coons, A.H., H.J. Creech, R.N. Jones and E. Bernner. 1942. The demonstration of pneumococcal antigen in tissues by the use of fluorescent antibody. *J Immunol, 45*:159.

Fantasia, L.D., W.H. Sperber, and R.H. Deibel: 1969. Comparison of two procedures for detection of *Salmonella* in food, feed and pharmaceutical products. *Appl Microbiol, 17*:540.

Fantasia, L.D.: 1969. Accelerated immunofluorescence procedure for the detection of *Salmonella* in foods and animal by-products. *Appl Microbiol, 18:*708.

Fantasia, L.D., J.P. Schrade, J.F. Yager, and D. Debler: 1975. Fluorescent antibody method for the detection of *Salmonella:* development, evaluation, and collaborative study. *J Assoc Off Anal Chem, 58:*828.

Georgala, D.L. and M. Boothroyde: 1964. A rapid immunofluorescence technique for detecting salmonellae in raw meat. *J Hyg, 62:*319.

Georgala, D.L., M. Boothroyde, and P.R. Hayes: 1965. Further evaluation of a rapid immunofluorescence technique for detecting salmonellae in meat and poultry. *J Appl Bacteriol, 27:*421.

Haglund, J.R., J.C. Ayres, A.M. Paton, A.A. Kraft, and L.Y. Quinn: 1964. Detection of *Salmonella* in eggs and egg products with fluorescent antibody. *Appl Microbiol, 12:*447.

Insulata, N.F., S.F. Schulte, and J.H. Berman: 1967. Immunofluorescence technique for detection of salmonellae in various foods. *Appl Microbiol, 15:*1145.

Munson, T.E., J.P. Schrade, N.B. Bisciello, Jr., L.D. Fantasia, W.H. Hartung, and J.J. O'Conner. 1976. Evaluation of an automated fluorescent antibody procedure for detection of *Salmonella* in foods and feeds. *Appl Environ Microbiol, 31:*514.

Olson, J.C., Jr.: 1972. *Bacteriological Analytical Manual,* Chapter VIII. FDA-DHEW, Washington, D.C.

Schulte, S.J., J.S. Witzemann, and W.M. Hall: 1968. Immunofluorescent screening for *Salmonella* in foods: comparison with culture methods. *J Assoc Off Anal Chem, 51:*1334.

Silliker, J.H., A. Schmall, and J. Chiu: 1966. The fluorescent antibody technique as a means of detecting salmonellae in foods. *J Foods Sci, 31:* 240.

Thomason, B.M., G.A. Hebert, and W.B. Cherry: 1975. Evaluation of a semi-automated system for direct fluorescent antibody detection of salmonellae. *Appl Microbiol, 30:*557.

Chapter 9

RAPID DETECTION OF WATER-BORNE FECAL COLIFORMS BY $^{14}CO_2$ RELEASE

D. J. REASONER AND E. E. GELDREICH

INTRODUCTION

IN RECENT years, particularly during the last decade, tremendous advances have been made in scientific instrumentation as well as in data storage and retrieval systems. As a result, new techniques and applications have been developed in many scientific areas including microbiology, and considerable impetus has developed toward rapid and/or automated microbiological techniques. Evidence of this is indicated in the attendance at this conference and at other recent meetings or conferences such as those held in London (Baillie and Gilbert, 1970), Stockholm (Hedén et al., 1975a, 1975b), and Tucson, Arizona (Rapid Diagnostic Techniques in Clinical Microbiology. ASM Conference, December 4-6, 1974, Tucson, Arizona, U.S.A.)

Although considerable progress in rapid or automated methods has been made in the fields of medical microbiology and immunology, and to some extent in bacterial characterization and identification, similar developments in water bacteriology have been slow to materialize. In sanitary bacteriology few changes have been made in the methodology used to detect or quantitate coliform bacteria. Current methods using either the membrane filter or the most probable number techniques yield results in a minimum of 24 to 48 h. Thus, in the examination of drinking water, contaminated water may already be consumed before public health officials become aware of the contamination and could take corrective action.

As a further area of concern, recent evidence that some organic compounds in finished water can react with chlorine to form carcinogenic compounds has stimulated a trend toward changing potable water treatment practices. Changing the treatment technology may alter the protective barrier inherent in maintaining a free chlorine residual throughout the distribution system. If this protective barrier is thus reduced, the need for rapid bacteriological monitoring systems and for emergency treatment technology will be mandatory.

Therefore, a rapid method for coliform detection is needed. In addition, the need exists for the development of a semi– or fully automated rapid coliform detection system for in-line use in water treatment plants or at strategic monitoring points in the distribution system. Such a system should be able to provide bacteriological results within 6 h or less and indicate the presence of a potential public health problem caused by a breakdown in the treatment process. Remedial measures could then be initiated up to 18 h sooner than is currently possible.

The idea of using ^{14}C-labeled compounds to aid in detection, quantitation, or identification of indicator bacteria in water is not new. Interest in the use of radioactively tagged compounds to aid in the detection of coliform bacteria in water stems from early work by Levin et al. (1956a, 1956b). They reported that presumptive evidence of coliforms in water could be obtained in 1 h using lactose broth containing ^{14}C-1-lactose and 37°C incubation. Theoretically, less than twenty cells of *Escherichia coli* could be detected after 1 h incubation. Later, Levin et al. (1957) found that the concentration of radioactive lactose had a direct effect on $^{14}CO_2$ production by *E. coli*, as did the pH of the medium. Maximum $^{14}CO_2$ production after 1 h at 37°C occurred at pH 8.0, but much of the $^{14}CO_2$ remained in solution. They also noted that nonmetabolic $^{14}CO_2$ release was a problem.

In 1959, Levin et al. described a modification of their procedure in which ^{14}C-sodium formate (0.01 percent) was used instead of ^{14}C-lactose and the incubation period at 37°C was increased to 4 h. However, the results obtained for ^{14}C-sodium formate did not correlate well with the results for ^{14}C-lactose. The main reason for this may have been the fact that the concentration of ^{14}C-sodium formate was fifty times less than the minimum concentration of ^{14}C-lactose (0.5 percent) (Levin et al., 1957).

The radioisotope procedure was further modified by Levin et al. (1961) to involve membrane filtration, ^{14}C-sodium formate in MacConkey broth and incubation at 44°C for 4 h. Their results showed that the lag period before ^{14}CO$_2$ was produced was longer for membrane filtered coliforms than it was for nonmembrane filtered coliforms; and ^{14}CO$_2$ production after 4 h by membrane filtered cells was 10 to 30 fold less than for nonfiltered cells. These effects were observed even when membrane filters were added to MacConkey broth containing nonmembrane filtered cells. Further, the adverse effect of the membrane filter on either cell multiplication or ^{14}CO$_2$ production from ^{14}C-sodium formate was also observed for *Pseudomonas* sp. and *Alcaligenes* sp. The filters had no effect on nonmetabolic ^{14}CO$_2$ from uninoculated controls. Additional tests showed that the adverse effect of filters on ^{14}CO$_2$ evolution varied significantly among brands of filters; the authors were unable to determine the cause of these effects.

Studies by Pijck and Defalque (1963) were in general agreement with those published by Levin et al. (1961). Pijck and Defalque used ^{14}C-sodium formate as the substrate for detection of *E. coli* in water. With a large coliform inoculum in pure culture experiments a positive test could be obtained after 10 min of incubation. Small inocula of *E. coli* took considerably longer to detect although thirteen cells of *E. coli* could be detected in 1 h. In polluted water samples, 4.1×10^3 cells could be detected in 2 h. Pijck and Defalque suggested the possible application of the ^{14}C-radioisotope technique as an automated method for the examination of potable water.

Scott et al. (1964a, 1964b) recommended the use of a ^{14}C-radiometric test for the detection of coliforms in potable water, raw surface water, and sewage. They used 1.17 microcuries of ^{14}C-sodium formate per test and a total incubation time of 4 h at 35°C. They concluded from their results that the collected ^{14}CO$_2$ radioactivity of ten or more counts per min (CPM) above the background CPM was an indication of the presence of coliforms or of a high noncoliform bacterial population in the water. Levin et al. (1962) showed that a number of soil bacteria can metabolize a mixture of ^{14}C-glucose (uniformly labeled) and ^{14}C-sodium formate and release $^{14}CO_2$ within a 3.5 h period. Water samples containing large numbers of some types of soil bacteria could, therefore, interfere with the ^{14}C-radiometric test for coliforms. In most cases of potable water examination, this would probably not be a problem.

Korsh et al. (1968), using a procedure similar to that of Scott et al. (1964a, 1964b) except for the incubation temperature (42°C instead of 35°C), found a direct correlation between the number of *E. coli* present and the $^{14}CO_2$ released from the ^{14}C-sodium formate. However, the results were not always consistent and problems were encountered due to a high level release of non-metabolic $^{14}CO_2$ from the ^{14}C-sodium formate. This release was attributed to redox reactions in the medium, disruption of the sodium formate chemical bonds by the beta radiation, or pH.

Korsh et al. (1971) used ^{14}C-1, 6-glucose in rosolic acid broth and Endo broth to detect *E. coli* in water. *E. coli* cultures isolated from water were used to prepare calibration graphs for correlation of the evolved $^{14}CO_2$ with the number of *E. coli* present. Incubation was at 42°C for 6 h. The authors reported that ten cells of *E. coli* per sample could be detected using Endo broth, while 30 to 40 cells per sample could be detected using rosolic acid broth. The amount of metabolic $^{14}CO_2$ released by *E. coli* in effluent samples was about 1.5 times less than was predicted from the calibration graph prepared. Korsh et al. suggested that the discrepancy could be explained by the inaccuracy of determining the actual *E. coli* count of the effluent sample.

Khanna (1973) used a radiometric procedure in which bacterial cells, exposed to phosphorous 32 (^{32}P) during 37°C incubation for various time periods, were coprecipitated with excess ^{32}P using egg white. After washing to remove the ^{32}P not taken up by the bacterial cells, the precipitate was filtered (Whatman No. 40® filter paper disks) and dried, and its ^{32}P-activity determined using a Geiger-Mueller counter. The results reported were obtained primarily from pure culture studies. Data presented for the examination of field samples showed that the coprecipitation and the membrane filter tests were reproducible. However, the results did not provide a calibration graph whereby *E. coli* or coliform density could be predicted or estimated from the ^{32}P-coprecipitation test. Also, the data did not provide a good indication of the sensitivity of the procedure for determining the coliform density of surface water samples, compared to the standard coliform procedure.

More recently, Bachrach and Bachrach (1974) used *E. coli* suspensions to conduct calibration and detection experiments with lactose nutrient broth containing 0.1 microcurie (μCi) of ^{14}C-lactose. Detection times for 1 to 10, 100, and 1,000 cells of *E. coli* were 6, 3, and 2 h, respectively. The procedure was not used to detect coliform bacteria in natural water samples, although aged suspensions of *E. coli* in water caused the release of [14]CO_2 within 1 h at 37°C.

Table 9-I summarizes the media, ^{14}C-substrate concentrations, radioactivity per test (microcuries), and mean [14]CO_2 CPM per bacterial cell reported by various investigators for rapid radiometric detection of coliforms in water. The mean [14]CO_2 CPM per bacterial cell was calculated from data in the cited references. The wide range of both ^{14}C-substrate concentration and [14]CO_2 activity evolved per bacterial cell illustrates two of the problems encountered in developing a good radiometric procedure. These are specifically the problems of standardizing both the type and concentration of labeled substrate and establishing the test conditions that stabilize the [14]CO_2 evolved per cell.

Since the basic principle of this procedure offers a promising approach to the urgent need for rapid monitoring of potable water quality, the technique is being explored in the authors' laboratory. Investigations by the authors have shown the need to further define and control variables so that the desired sensitivity of one coliform per 100 ml may be obtained along with better precision of test replication.

The authors' research has been directed toward detection of labeled carbon dioxide by liquid scintillation spectrometry in order to take advantage of the increased counting efficiency and sensitivity of this method. In addition, the authors have also been investigating variations in incubation temperature and comparing

TABLE 9-I

SUMMARY OF MEDIA AND LABELED SUBSTRATES USED IN REPORTED
^{14}C-RADIOMETRIC TESTS FOR COLIFORM BACTERIA IN WATER

Investigators	Medium	^{14}C-substrate concentration[a]	^{14}C, μCi per test[b]	4-hr mean $^{14}CO_2$ cpm/cell[c]
Levin et al. (1956)	Lactose Broth	^{14}C-1-lactose, 0.5%	89.5	11.85-44.3
Levin et al. (1957)	Lactose Broth	^{14}C-1-lactose, 0.3%	7.2	0.04-0.42
Levin et al. (1959)	Brilliant Green Lactose Bile Broth	^{14}C-sodium formate, 0.01%	. . .	0.49-0.98
Levin et al. (1961)	MacConkey MF Broth (British)	^{14}C-sodium formate, 0.002%	1.15	6.77
Scott et al. (1964a,b)	m-Endo MF Broth	^{14}C-sodium formate, 0.008%	1.17	1.0-2.0
Korsh et al. (1968)	Hottinger's Medium	^{14}C-sodium formate	0.12-0.24	0.04
Korsh et al. (1971)	Rosolic Acid Broth	^{14}C-1, 6, -glucose	0.2	0.5
	Endo Broth	^{14}C-1, 6, -glucose	0.2	0.56-1.72
Bachrach and Bachrach (1974)	Nutrient Broth plus 2 x 10^{-5}M lactose	^{14}C-lactose	0.1	0.0002-0.08

a. Percent ^{14}C-substrate given by investigators; in some reports, ^{14}C-substrate concentration was not given and could not be calculated from the data.

b. ^{14}C-substrate activity in microcuries (μCi) as reported by the investigators; in reports where the ^{14}C-substrate concentration was not stated, the value was calculated from the data in the report.

c. The ^{14}CO_2 activity per cell (cpm/cell), if not stated, was calculated on the basis of the cpm evolved from the bacterial concentration closest to 100 cells/ml during a four-hr. incubation period.

the efficiency of ^{14}C-lactose (labeled on one carbon only), ^{14}C-glucose, and ^{14}C-mannitol as substrates for a rapid fecal coliform test.

MATERIALS AND METHODS

The culture medium used was standard methods m-FC broth (APHA, 1971), distributed in 5 ml volumes to 118 ml (4 oz.) wide mouth jars. The jar lids were black bakelite with an inside diameter of approximately 5 cm, large enough to accept a 4.25 cm diameter disc of Whatman No. 1® filter paper.

Appropriate volumes or dilutions of surface water samples were filtered through 47 mm diameter membrane filters, 0.47 μm pore size. After filtration, each membrane filter was cut in half with sterile scissors and placed in a jar containing 5 ml of m-FC broth. The jars were then capped and placed in a shaking water-bath at 35°C for 2 h. Then 1 ml of m-FC broth containing 0.25 ml (1.2 microcuries radioactivity) of uniformly labeled ^{14}C-glucose, uniformly labeled ^{14}C-mannitol, or ^{14}C-lactose (1C-glucose) was added to each sample jar. At the same time, a CO_2 trapping system, consisting of a 4.25 cm diameter disc of Whatman No. 1 filter paper moistened with 0.25 ml of a saturated barium hydroxide solution, was placed in each jar lid and the lids placed on the jars. The jars were again placed in the water-bath and the temperature raised to 44.5°C. The time required for the temperature to change from 35°C to 44.5°C was about 10 min. The cultures were then incubated for an additional 2 to 8 h. For most experiments, the total incubation time after inoculation was generally 4 to 6.5 h, depending upon the experiment. Samples were removed for $^{14}CO_2$ assay at 0.5 h or hourly intervals by re-placing each jar lid with another lid containing a new filter paper disc freshly moistened with the saturated barium hydroxide solution.

The $^{14}CO_2$ was trapped by precipitation as $Ba^{14}CO_3$ on the $Ba(OH)_2$ moistened filter paper discs. The discs were dried and placed individually in appropriately marked liquid scintillation vials. Five ml of liquid scintillation cocktail were added to each vial and the vials then capped and placed in a Packard Tri-Carb Liquid Scintillation Spectrometer (Model 3320)® for assay. The $^{14}CO_2$ activity (in counts per min, CPM) of each sample was corrected for control activity and plotted as unaccumulated or accumulated values. The results were correlated with the initial coliform count, determined by standard methods procedures.

RESULTS

Figure 9-1 compares fecal coliform release of $^{14}CO_2$ from uniformly labeled ^{14}C-glucose and from ^{14}C-lactose in 8 h parallel experiments using a starting inoculum of 35 and 350 cells. The data shows that release of $^{14}CO_2$ from labeled lactose was con-siderably delayed compared to release from uniformly labeled glucose: 4.5 to 5 h for the 35 cell inoculum and 3 h for the 350

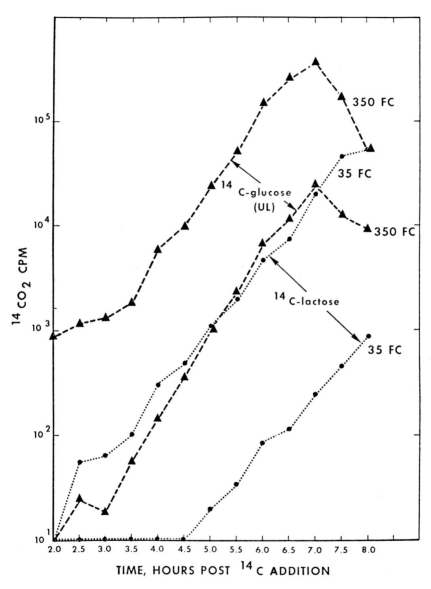

Figure 9-1. Comparison of $^{14}CO_2$ counts per min (CPM) released from ^{14}C-glucose (UL) and ^{14}C-lactose (1C-glucose). 35 and 350 fecal coliform cells per test. Data plotted are corrected, unaccumulated $^{14}CO_2$ CPM.

cell inoculum. Also, the overall release of $^{14}CO_2$ from the labeled lactose was lower than that from uniformly labeled glucose. This may have been due partly to the differences in the millimolar concentrations of the two substrates. Also, the only labeled carbon in the lactose was at the number one position of the glucose moiety, while all six carbons of the d-glucose were labeled. For both substrates, 1.2 microcuries of activity were added to the m-FC broth to keep the total radioactivity at the same level.

Figure 9-2 shows averaged, normalized data from three experiments in which uniformly labeled ^{14}C-glucose was used as the substrate, added after 2 h incubation at 35°C. Following addition, the reaction mixtures were incubated at 44.5°C for 5 h. The data points represent $^{14}CO_2$ counts per min corrected for control activity. The results show that with an initial inoculum ranging from 2 to 2,000 cells, the minimum detectable fecal foliform limit appeared to be about 20 cells. The curves do not parallel each other, although the curves for 200 and 2,000 cells are nearly parallel. Each tenfold difference in the number of cells present did not result in a tenfold difference in the $^{14}CO_2$ CPM released.

Figure 9-3 illustrates results from experiments in which uniformly labeled ^{14}C-mannitol was used as the substrate. These experiments and those shown in Figure 9-2 were conducted in parallel and the results handled in the same way. The data indicate that there was no overlap in the collected $^{14}CO_2$ activity at any time interval for any of the initial inoculum levels. The curves for the 20, 200, and 2,000 cells were approximately parallel and there was generally a tenfold increase in $^{14}CO_2$ CPM for each tenfold increase in cell concentration. The lower sensitivity limit appeared to be between 2 and 20 cells.

Figure 9-4 shows labeled CO_2 (CPM) released from ^{14}C-glucose at different time intervals; Figure 9-5 shows similar results for ^{14}C-mannitol. Comparison of the curves illustrates that the labeled mannitol was a better substrate for $^{14}CO_2$ release than labeled glucose, because the correlation with the fecal coliform inoculum was more nearly directly proportional throughout the experiment. Also the results after 2 h and 2.5 h appeared more reliable for mannitol than for glucose.

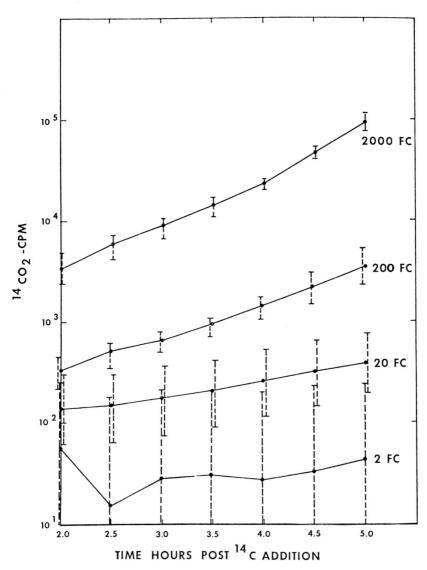

Figure 9-2. $^{14}CO_2$ released from ^{14}C-glucose (UL) by 2, 20, 200, and 2,000 cells of fecal coliforms (FC) per test. Data plotted are normalized, corrected, accumulated $^{14}CO_2$ CPM.

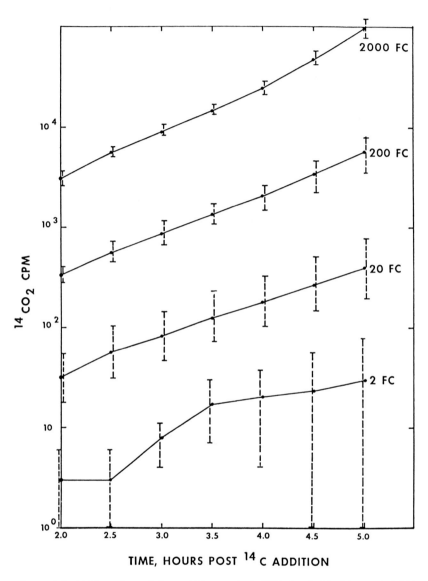

Figure 9-3. $^{14}CO_2$ released from ^{14}C-mannitol (UL) by 2, 20, 200, and 2,000 cells of fecal coliforms (FC) per test. Data plotted are normalized, corrected, accumulated $^{14}CO_2$ CPM.

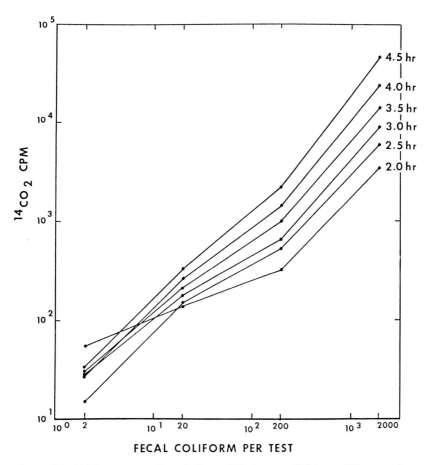

Figure 9-4. $^{14}CO_2$ CPM released from ^{14}C-glucose (UL) at different time intervals. Data plotted are normalized, corrected, accumulated $^{14}CO_2$ CPM.

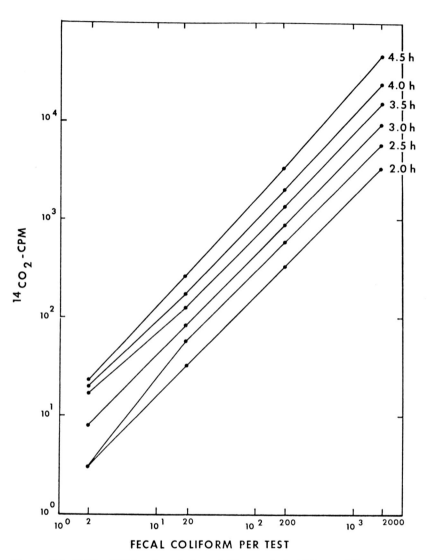

Figure 9-5. $^{14}CO_2$ CPM released from ^{14}C-mannitol (UL) at different time intervals. Data plotted are normalized, corrected, accumulated $^{14}CO_2$ CPM.

DISCUSSION

The procedure of Scott et al. (1964a, 1964b) used ^{14}C-sodium formate as the substrate and detection of the labeled CO_2 was accomplished using a gas-flow proportional counter. The authors' attempt to convert their procedure to one that uses liquid scintillation spectrometry to detect $^{14}CO_2$ was only partially successful. Sodium formate proved to be very unsatisfactory as a labeled substrate due to the high background of nonmetabolic $^{14}CO_2$ CPM obtained from uninoculated controls and also to the increased sensitivity of the liquid scintillation technique. Therefore, experiments using uniformly labeled ^{14}C-glucose and ^{14}C-mannitol and ^{14}C-lactose were initiated. Early experiments showed that agitation during incubation improved results, a finding reported by other workers (Deland and Wagner, 1969; De Blanc et al., 1971). While ^{14}C-lactose would seem to be the substrate of choice based on the accepted coliform definition and methodology presently used, reliable results were not obtained in the desired time frame of 4 to 5 h. Cost is another consideration, since ^{14}C-lactose is about three times as expensive as ^{14}C-glucose or ^{14}C-mannitol. Uniformly labeled ^{14}C-lactose is currently being synthesized for use by the authors, but only for comparative purposes since it is prohibitively expensive.

The use of ^{14}C-glucose (UL) presents a problem because many bacteria in water can metabolize this sugar and thus make it unavailable for use by the coliforms. Therefore, the standard lactose-based m-FC broth was chosen for continued use, but the procedure included a 2 h resuscitation period at 35°C followed by addition of ^{14}C-substrate and elevation of the incubation temperature to 44.5°C. In this way, the selective advantages of the lactose and inhibitory agents in the medium were retained. The resuscitation period permits the fecal coliform cells to repair metabolically and to begin metabolizing before the ^{14}C-labeled substrate is added. Upon addition of the ^{14}C-labeled sugar, the coliforms quickly outgrow and out-metabolize the noncoliform bacteria. Additionally, the change to 44.5°C incubation assures selection for the fecal coliform population. The results obtained indicate that this procedure is effective.

Experiments using uniformly labeled mannitol were prompted by an earlier report (Pugsley et al., 1973) which indicated that mannitol could be used in place of lactose in the 44°C confirmatory test for *E. coli* (fecal coliforms). The results thus far indicate that ^{14}C-mannitol is a suitable fermentation substrate and yields better results than ^{14}C-glucose or ^{14}C-lactose in this rapid procedure. In addition, the background of nonmetabolic [14]CO$_2$ CPM from the ^{14}C-mannitol was consistently lower than for any of the other ^{14}C-substrates. Also the correlation between fecal coliform concentration and the [14]CO$_2$ CPM was much better using ^{14}C-mannitol than with the other labeled substrates.

One of the problems encountered with this test is that although the reproducibility of results from replicates run simultaneously was good with both ^{14}C-glucose or ^{14}C-mannitol, tests on water samples collected on different days but containing equivalent fecal coliform concentrations may produce results that differ widely — a problem that was also noted by Levin et al. (1961). This may be related to the relative recentness of the pollution and thus to the degree of stress to which the coliforms have been subjected; the longer coliforms are in water, the greater the stress and the greater the possibility that they will yield a relatively low amount of [14]CO$_2$. Consequently, continued development of this test will be directed toward controlling this problem and establishing the specificity of the test.

The 4 to 5 h sensitivity level of the test using labeled mannitol appears to be between two and twenty cells. Bachrach and Bachrach (1974) reported that one to ten cells could be detected using ^{14}C-lactose, but they used pure cultures, not natural water samples. Using ^{14}C-1-6-glucose and 42°C incubation, Korsh et al. (1971) reported detection of ten cells of *E. coli* per sample with Endo medium and thirty to forty cells with rosolic acid medium. Again, pure cultures were used, although one set of natural water samples was found to show reasonably good correlation with pure culture results.

Potential for Semiautomated or Automated Procedure

Assuming that the specificity, sensitivity, and reliability of this method prove to be acceptable, future research will explore the possibility of the development of a semiautomated or perhaps fully automated procedure. In a fully automated system the instrument would be programmable and could be set up to take samples at specified time intervals during a 24 h period. In this way, fixed interval surveillance of a water supply could be maintained even though samples would not be taken continuously. In a semiautomated procedure certain specific steps would be handled manually.

A semiautomated procedure would probably be the most suitable for several reasons. (a) Membrane filtration requires several handling steps that might prove difficult to adapt to automation. Also, at this point the technician would have greater control over the test in terms of sample volume and membrane replacement, in case of membrane rupture and the need to resample. (b) Dual temperature incubation requires a sequential move of the sample chamber from one temperature to another. (c) Addition of the ^{14}C-substrate would need to be keyed to the move of each sample chamber from $35\,°C$ to $44.5\,°C$ incubation. (d) Detection of $^{14}CO_2$ requires physical removal of the $Ba^{14}CO_3$ precipitate-filter disc from the incubation chamber and transfer to the liquid scintillation vial, or purging of the sample chamber head-space gas to collect the labeled CO_2.

Buddemeyer (1974) developed a two-compartment vial for cumulative and continuous measurement of *in vitro* bacterial metabolism using radioactively labeled substrates. This or other such devices may be adaptable for use in a semiautomated or automated procedure. For detection of $^{14}CO_2$ by means of purging the gas into a detector, there is a commercially available instrument (Bactec®, Johnston Laboratories, Inc., Cockeysville, Md.) that, with modification, might prove suitable for either semiautomated or automated use. Preliminary experiments with the current manual instrument indicate that the liquid scintillation spectrometer is more sensitive than the Bactec instrument.

Table 9-II shows the various steps of the labeled CO_2 method

TABLE 9-II.

STEPS IN FECAL COLIFORM $^{14}CO_2$ DETECTION PROCEDURE —
SUITABILITY FOR SEMIAUTOMATED OR AUTOMATED PROCEDURE

Step	*Semiautomated*	*Automated*
SAMPLE HANDLING		
Collection	Manual	Yes
Filtration	Manual	Yes
Sample Inoculation	Manual	Yes
INCUBATION		
35°C — 2 h.	Yes	Yes
^{14}C-Substrate Addition	Manual	Yes
44.5° — 2 to 2.5 h.	Yes	Yes
$^{14}CO_2$ HANDLING		
Trapping		
Precip. as $BA^{14}CO_3$	Yes	Yes
Purge to Ionization		
Chamber, Electrometer	Manual; Yes	Yes
Detection		
Liquid Scintillation		
Counting	Manual; Yes	Yes
Ionization Chamber		
Electrometer	Manual; Yes	Yes
RESULTS		
Data Printout	No; Yes	Yes
Interpretation	Operator	Yes; Go/No Go
SAMPLE DISPOSAL		
Sterilize	Manual; Yes	Yes
Discard	Manual; Yes	Yes

and indicates the potential of each step for automation. The difference between a semiautomated and an automated instrument might only be that in the former case a technician would have to manually advance the samples through specific steps of the procedure, while in the latter case all steps would be fully programmed. A fully programmed instrument would probably permit the choice of selected or random sampling and fixed-interval sampling and analysis.

Figure 9-6 is a schematic representation of how the procedure might be automated; a technician would only maintain the supply of ^{14}C-labeled medium, check the instrument periodically to assure proper function, and insert positive and negative controls on a daily basis. Continuous fixed-interval water quality monitor-

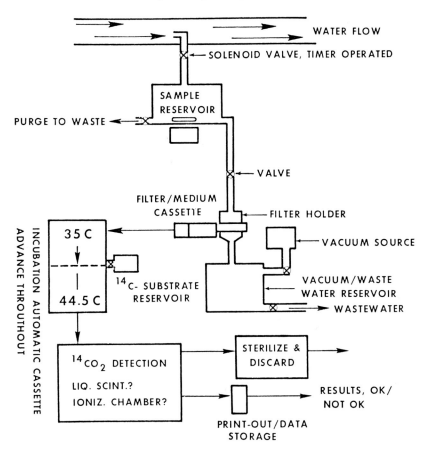

Figure 9-6. Schematic outline of automated fecal coliform $^{14}CO_2$ procedure.

ing in large municipal treatment plants would probably require such an automated system. The manpower required to staff it would be minimal compared to the staff that would be needed to maintain 24 h surveillance using manual procedures.

The potential usefulness of this method as a manual, semi-automated, or a fully automated procedure can be extended to the examination of water samples from bathing and recreational waters, stream pollution, and emergency water examination. Thus, the method has considerable merit and potentially far-reaching applications.

REFERENCES

American Public Health Association. 1971. *Standard Methods for the Examination of Water and Wastewater,* 13th ed. American Public Health Association, Inc., New York.

Bachrach, U. and Z. Bachrach: 1974. Radiometric method for the detection of coliform organisms in water. *Appl Microbiol, 28:*169.

Baille, A. and R.J. Gilbert (eds): 1970. *Automation, Mechanization and Data Handling in Microbiology.* Society for Applied Bacteriology, Technical Series No. 4. Acad Pr, New York.

Buddemeyer, E.U.: 1974. Liquid scintillation vial for cumulative and continuous radiometric measurement of *in vitro* metabolism. *Appl Microbiol, 28:*177.

DeBlanc, H.J., Jr., F.H. Deland, and H.N. Wagner, Jr.: 1971. Automated radiometric detection of bacteria in 2,967 blood cultures. *Appl Microbiol, 22:*846.

Deland, F.H. and H.N. Wagner: 1969. Early detection of bacterial growth, with carbon-14-labeled glucose. *Radiology, 92:*154.

Hedén, C. and T. Illéni (eds..): 1975. *Automation in Microbiology and Immunology.* Symposium on Rapid Methods and Automation in Microbiology. Stockholm, Sweden, June 3-9, 1973. Sponsored by UNESCO, WHO, and the International Organization for Biotechnology and Bioengineering. Wiley, New York.

Hedén, C. and T. Illéni (eds.): 1975. *New Approaches to the Identification of Microorganisms.* Symposium on Rapid Methods and Automation in Microbiology, Stockholm, Sweden, June 3-9, 1973. Sponsored by UNESCO, WHO, and the International Organization for Biotechnology and Bioengineering. Wiley, New York.

Khanna, P.: 1973. Enumeration and differentiation of water bacteria with phosphorous 32. *J Water Poll Contr Fed, 45:*262.

Korsh, L.E., V.F. Zheverzheeva, and S.G. Egorova: 1968. Concerning the rapid detection of *E. coli* in water using radioisotope C^{14}. *Hyg Sanit* (Trans. from *Gig Sanit), 33:*375.

Korsh, L.E., O.I. Yurasova, A.G. Nikonova, and M.A. Motova: 1971. Use of ^{14}C for rapid *E. coli* counts in water. *Gig Sanit, 36:*78.

Levin, G.V., V.R. Harrison, and W.C. Hess: 1956. Preliminary report on a one-hour presumptive test for coliform organisms. *J Am Water Works Assoc, 48:*75.

Levin, G.V., V.R. Harrison, W.C. Hess, and H.C. Gurney: 1956. A radioisotope technic for the rapid detection of coliform organisms. *Am J Public Health, 46:*1405.

Levin, G.V., V.R. Harrison, and W.C. Hess: 1957. Use of radioactive culture media. *J Am Water Works Assoc, 49:*1069.

Levin, G.V., V.R. Harrison, W.C. Hess, A.H. Heim, and V.L. Strauss: 1959.

Rapid, radioactive test for coliform organisms. *J Am Water Works Assoc, 51*:101.

Levin, G.V., V.L. Strauss, and W.C. Hess: 1961. Rapid coliform organisms determination with ^{14}C. *J Water Poll Control Fed, 33*:1021.

Levin, G.V., A.H. Heim, J.R. Clendenning, and M.F. Thompson: 1962. "Gulliver"—a quest for life on Mars. *Science, 138*:114.

Pijck, J. and A. Defalque: 1963. Le dépistage précoce de micro-organismes coliformes par voie radiometrique. *J Pharm Belg, 18*:180.

Pugsley, A.P., L.M. Evison, and A. James: 1973. A simple technique for the differentiation of *Escherichia coli* in water examination. *Water Research, 7*:1431.

Scott, R.M., D. Seiz, and H.J. Shaughnessy. 1964a. I. Rapid carbon[14] test for coliform bacteria in water. *Am J Public Health, 54*:827.

Scott, R.M., D. Seiz, and H.J. Shaughnessy: 1964b. II. Rapid carbon[14] test for sewage bacteria. *Am J Public Health, 54*:834.

Chapter 10

ENUMERATION OF BACTERIA USING HYDROPHOBIC GRID-MEMBRANE FILTERS*

A. N. SHARPE AND G. L. MICHAUD

Abstract

The hydrophobic grid-membrane filter (HGMF) consists of a grid of hydrophobia wax barriers on the surface of a conventional membrane filter. On conventional membrane filters microbial colonies grow in random positions and with a wide distribution of sizes; the error in count caused by overlapping of colonies cannot easily be determined, and there is a relatively low limit to the number of colonies which may be safely counted. The grid of the HGMF, however, predetermines the positions and sizes of colonies, and effectively separates them from one another, eliminating this uncertainty. The HGMF may be treated theoretically in much the same way as a Most Probable Number apparatus, operating at a single dilution but having a very large number of growth compartments (grid-cells). Because of the small size of the grid-cells very high numbers of growth units (GU) can be enumerated on one HGMF: square (60 mm) HGMFs containing 10,000 grid-cells are capable of enumerating up to 9×10^4 microorganisms in water or foods. Thus, the need for dilutions is reduced or eliminated. Use of HGMFs can lead to more accurate electronic colony counts than is presently possible because the signal to noise ratio is improved, colonies have uniform peak-widths, the need to recognize colonies by their edges is eliminated, and saturation of the HGMF does not lead to false low readings. The HGMF is an attractive subcomponent for various mechanized or automated microbiological systems.

---

*We are very much indebted to Mr. M. P. Diotte and Mr. Y. Dudas for their contributions to the electronic and mechanical engineering aspects of HGMFs, and to Mrs. G. R. Roberts and Dr. J. Hall for their contributions to the statistical and mathematical aspects.

INTRODUCTION

THE mechanization of any process must generally be approached from two directions. Consideration must be given to the design and suitability of process equipment and also to the possibility of modifying the process so as to reduce the problems encountered in its mechanization. This duality of approach certainly applies to microbiology; useful developments in microbiological mechanization will almost certainly be accompanied by, or even depend on, the development and acceptance of processes more suited to mechanical handling than are most of the analytical methods presently in use.

Some Problems of Mechanizing Microbiology

In conventional plate or membrane filter techniques, microbial cells are deposited randomly over the growth area. These cells then multiply to form colonies which vary in size, shape, optical density, and sharpness. This uncontrolled growth causes unnecessary labor for the microbiologist, but more importantly, it also presents great problems to the developer of mechanized or automated systems. For example, it reduces the resolving power of the individual plate, resulting in the need to prepare serial dilutions for each sample. In addition, the variability and poor signal characteristics of uncontrolled growth reduces the accuracy and value of electronic counters, thus hindering the introduction of such instruments and the complete automation of quantitative microbiology.

The number of colonies that may grow randomly in a given area before a serious loss of counts occurs as a result of overlapping depends on the average size of the colonies. Obviously, the larger the average size, the greater is the chance of two or more colonies being incorrectly counted as one, and therefore the fewer colonies that can be tolerated. In the extreme, if there are too many colonies of too large a size, the whole area will be covered by an unresolvable lawn from which no count can be obtained.

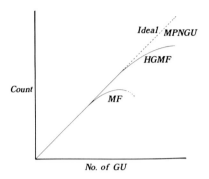

Figure 10-1. Conceptual relationship between observed "colony count," and number of growth units (GU) on a conventional membrane filter (MF), a hydrophobic grid-membrane filter (HGMF), and an ideal system capable of resolving every colony. The line HGMF describes the HGMF positive grid-cell count, which reaches a plateau, whereas the MPNGU value derived from this is indistinguishable from the ideal case.

The counting efficiency of conventional membrane filter (MF) or plating systems is shown in Figure 10-1. A system which had perfect recovery, i.e. which produced one resolvable colony from every growth unit (GU),* would yield a straight line of unit gradient. Conventional MF or plate methods, however, begin to deviate from linearity at quite a low number of growth units, soon reach a maximum, and then become meaningless as the "lawn condition" is approached. In theory, at least, one could apply a correction for the loss of counts caused by overlap if one had data on both the size distribution of the colonies and the technician's ability to resolve intersecting circles. In practice, this is not possible, and it is therefore necessary to set fairly arbitrary limits to the number of colonies allowable on a counted filter or plate. For plates the upper limit is often 300. For MFs it is often 80, or even 60 if colonies tend to be large. However, the statistics of random numbers demands a reasonably large number of colonies (say 20

---

*The term "growth unit" is preferred to the more usual "colony forming unit" when dealing with HGMFs, since the growths on this apparatus are not always colonies, in the strictest sense. At low inoculum concentrations, growths on the HGMF are more likely to be pure cultures than on an MF; however, at high inoculum concentrations, although counting is still possible, several growth units may contribute to the growth in any particular positive grid-cell.

or 30) if the count is to be statistically significant; to obtain plates or MFs having colony numbers between these upper and lower limits several serial dilutions are often required. The dilution step can be mechanized, but it is a fairly formidable engineering problem, and the machinery involved tends to be complicated. The need to dilute is a definite obstacle to the development and introduction of automation in quantitative microbiology.

Two other problems, both also resulting from the existence of random colonies, must be overcome or avoided if conventional methodology is to be mechanized. The first is the development of a counting instrument capable of accurately counting colonies in the presence of other colonies, food debris, or other optical noise. The second is the development of a means of distinguishing reliably between cases where the number of colonies is small, and where it is large to the point of confluence. These are difficult problems. To build the required degree of performance, similar to that of the human eye, would be prohibitively expensive at present. This comment introduces a last major consideration, which is that microbiology is not a field in which large fortunes are likely to be made. The commercial attractiveness of very expensive, albeit highly productive, machines is thus small, so that equipment for microbiology must generally be relatively inexpensive, i.e. unsophisticated.

Faced with these difficult engineering problems, it is reasonable to look very closely at the second approach to mechanization — modification of the process so as to reduce the problems encountered. The remainder of this chapter discusses one possible process modification, the apparatus associated with it, and the manner in which its use may effect these problems.

DISCUSSION

The Hydrophobic Grid-Membrane Filter (HGMF)

The use of the HGMF in microbiology demands a significant but probably acceptable deviation from accepted technique. One of the benefits appears to be a real possibility of reducing the problems of mechanization. An alternative justification of the HGMF is described in Chapter 3, whereby its ability to provide a better match between the information coded in the colonies and the information in the microorganisms may be conducive to a more accurate count. Here, however, its characteristics are approached from a more conventional microbiological viewpoint.

The HGMF is a conventional MF, on which is printed a grid of hydrophobic wax barriers, in a simple process using hot wax (Sharpe, 1975). The surface is thus subdivided into a large number of growth compartments (Figs. 10-2A, 10-2B) which are made considerably smaller than the area normally occupied by colonies. Typically, the line spacing is 0.020 in, or 0.5 mm (Fig. 10-2C). The hydrophobic barriers confine colony development to the grid-cell in which the growth unit falls, and multiplication occurs in a habitat that is relatively independent of those around it. An important consequence of printing the grid is that formation of a colony from any particular growth unit can only take place in one of a finite number of positions, whose locations are predetermined and easily specified by simple coordinates. Two other consequences are also important: because of the restraint on lateral growth, much higher colony densities can be achieved

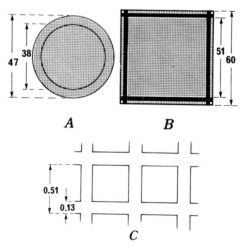

A   B

C

Figure 10-2. *A* and *B* depict relative sizes of inoculated areas of typical HGMFs. *C* illustrates typical HGMF barrier and growth area dimensions. All dimensions are in mm.

and, also, the colonies are forced to grow upwards so that they have a higher optical density. It is mainly through its effect on these colony growth characteristics that the HGMF appears to offer an easier means of mechanization than the conventional counting techniques.

The Dilution Requirement

One of the first problems of mechanization, as described above, is the need to prepare serial dilutions of each sample suspension. This problem can be minimized or eliminated by raising the upper limit of countable numbers of colonies on the filter or plate by two or three orders of magnitude, an advance that is realised, albeit indirectly, in the performance of the HGMF. One must regard the HGMF as a Most Probable Number device, because each grid-cell has a certain probability of being, or not-being, inoculated, depending on the number of microorganisms in the liquid filtered. The following equation was derived by Sharpe and Michaud (1975) to describe the relation between the most probable number of growth units (MPNGU) and any positive grid-cell count (x) up to (N-1) for an HGMF of N grid-cells.

$$MPNGU = -N \ log_e \ \frac{(N-x)}{N}$$

From this relation the MPNGU corresponding to any count of positive grid-cells may be calculated, either by use of a table derived from the equation, or by use of a nomogram similar to Figure 10-3.

When only a small proportion of the grid-cells have been inoculated the HGMF provides an excellent means of isolating species in a mixed culture, since there is a high probability that even closely spaced growth units will find themselves separated by a barrier, whereas in a conventional system they might grow confluently *(see below)*. When a large proportion of the grid-cells are inoculated it is obvious that there is a significant probability of two or more growth units growing confluently in the same grid-cell. In fact, when (N-1) grid-cells are inoculated there may be an average of eight or nine growth units per grid-cell.

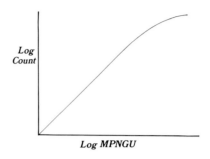

Figure 10-3. A nomogram such as this may be constructed from the equation $MPNGU = -N\,log_e\,\dfrac{(N-x)}{N}$ to facilitate calculation of MPNGU values.

The MPNGU value is, in effect, the same as the colony count on a conventional plate or filter, corrected for overlap losses in a way that is impossible with the conventional method. The correction is possible because the statistical treatment of the HGMF does not require data about the size distribution of the colonies.

The small size of the HGMF grid-cells, relative to the normal area occupied by colonies, allows them to be packed onto the filter at much greater densities, for example, 50 to 120 times greater, than is normally allowable for unrestrained growth. In addition, because the well-defined mathematical relation between count and numbers of growth units allows useful counts to be obtained when an average of seven or eight growth units are present in each grid-cell, the total density of growth units allowable on the HGMF may be up to 1,000 times greater than on a conventional filter. The 60 mm square HGMF shown in Figure 10-2B contains 10,000 grid-cells and has a maximum MPNGU of 92,000 growth units (corresponding to 9,999 positive grid-cells).

Remembering that the usual method of preparing microbiological sample suspensions by blending involves a decimal dilution of the sample, it can be seen that one HGMF is potentially able to enumerate samples containing up to 9×10^5 organisms per gram without the need for serial dilution. This limit is suitable for a very large number of foods, swabs, etc. In the case of water samples, the limit corresponds to 9×10^4 organisms per 100 ml, which is sufficient to cover the majority of water quality analyses. Thus, the use of the HGMF as a subcomponent in mechanized microbiological systems would reduce or eliminate the need for serial dilution and, in many situations, would allow a corresponding reduction in the complexity (and cost) of the instrument.

The practical performance of HGMFs with pure cultures of bacteria and samples of natural waters has been described by Sharpe and Michaud (1975). The HGMF is used in exactly the same manner as a conventional MF. The grid-cell count after incubation, however, is converted to MPNGU using tables, and this value is used as the measure of microbial contamination. Circular (47 mm) HGMFs containing 3,650 grid-cells gave linear recoveries of bacteria in nearly all cases, up to the theoretical maximum (saturation point) of 30,000 growth units. In practice, therefore, the HGMF mimics the performance of the perfect recovery system shown in Figure 10-1, even though it does not really resolve individual colonies as one approaches its upper limit.

A 47 mm circle is not the most efficient format for the HGMF, partly because there is always an uncertainty about the true number of grid-cells inoculated at the periphery of the filtration area and partly because it does not fully express the economy and organization of the coordinate reference system embodied in the grid. Our latest experiments have used the 100×100 grid-cell HGMF shown in Figure 10-2B. This filter has a wide printed border, slightly narrower than the square filtration funnel with which it is used, so that the area is delineated by the print rather than the funnel. Thus, filtration of equal volumes through all grid-cells is assured. The maximum number of GU which may be counted on such a HGMF is 9.2×10^4. Typical recovery graphs obtained for representative organisms are shown in Figure 10-4. The graphs are linear up to the theoretical saturation point of 92,000 GU. A slightly improbable example may be used to illustrate the performance of this HGMF: To enumerate cells of fecal coliforms at a concentration of 43,000 per inoculum, using the American Public Health Association (1971) recommended method with M-FC medium, would require six dilutions with seven platings of membrane filters. A single square HGMF would suffice.

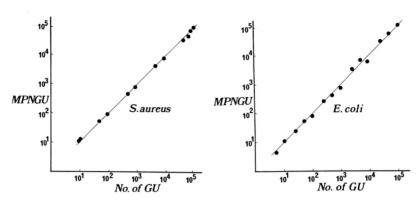

Figure 10-4. Recovery of *Escherichia coli* and *Staphylococcus aureus* on 10,000 grid-cell square HGMFs, illustrating linearity of MPNGU values up to 9 x 10⁴ GU. Filtration was effected using a square filtration funnel slightly larger than the printed HGMF border. Other experimental conditions were as described by Sharpe and Michaud (1975).

The Colony Counting Requirement

There are really two problems involved in the accurate counting of conventional colonies. First, there is the need for the instrument to recognize the boundaries between colonies and substrate, i.e. to be able to recognize colonies against optical noise in the plate. Second, there is the need for it to be able to distinguish between just a few colonies and a confluent lawn of growth. Until this latter capability is assured it is obviously impossible to automate the microbiological analysis, i.e. to leave everything to the machine.

The existence of these problems is simply due to the randomness of colony locations, sizes, and shapes, in a conventional plate or membrane filter count. From the point of view of a machine with less-than-human capability, this randomness obliges it to examine, i.e. scan, the whole of the growth area in order to detect any changes in optical density possibly representing colony boundaries. Unfortunately, of course, the observed changes may also be due to optical noise such as food debris, bubbles, ripples, etc., or they may not occur at all, as in the case of the lawn growth. The electronic counter has no prior information on the whereabouts of colonies, and thus must examine not only the areas occupied by colonies, but also those not occupied by colonies. The latter is generally a much greater area. For example, 300 circular colonies of 1 mm diameter occupy only 4 percent of the area of a 90 mm Petri dish. The unoccupied area, 96 percent, is thus fourteen times greater than that occupied by colonies. Unless the signal to noise ratio is very good, there will be a high probability of erroneous counts arising from the unoccupied areas.

Regardless of whether the actual scanning is done by television camera, flying spot, laser, or other scanners, a minimum resolving power is required to detect the smallest colonies. For a typical television camera of 525 line vertical resolution and horizontal resolution equivalent to 625 lines, the image can be regarded as being composed of approximately 330,000 picture points (the effect of interlacing half-frames of a normal television scan can be ignored here). For a circular growth area just filling the vertical frame height, approximately 216,000 of these picture points can be considered as contributing to the actual count. Thus, during a single complete scan the colony counter must make 216,000 "decisions" as to the presence or absence of a colony, based on a point-to-point comparison of the optical properties of the scanned area.

If the maximum allowable error is reasonably placed at no greater than 10 percent on a count of 300, i.e. no more than thirty wrong decisions per scan, the probability of making an erroneous decision must be less than one in 7,200. Unfortunately, the accuracy with which such decisions are made depends on the signal to noise ratio of the optical part of the system, and for most microbiological systems which contain food debris, translucent media, etc., the signal to noise ratio is poor. Thus the probability of making an error is high, and for this reason the available electronic colony counters perform badly in many practical situations of microbiological analysis. In the extreme case, where the colony concentration is high enough to produce a lawn, the number of detectable boundaries is small and a false low count is recorded.

The situation is much better with the HGMF or any other

Most Probable Number device, such as the microtiter plate. The location of any colony is now no longer random, but restricted to one of a relatively small number of predetermined locations, each of which can be inspected by the counter. It is possible to construct a counter which reads only the optical density (or other optical property) at the known center of each grid-cell, ignoring all other possible image points. In the case of the 47 mm circular HGMF there are only 3,650 decisions to be made; in the case of the 60 mm square HGMF there are 10,000 decisions. Both figures are very much smaller than the number of decisions required when colonies occur randomly. Thus, even if the accuracy with which colonies can be recognized is no better than at present, such a counter is likely to make fewer errors than conventional systems simply because it is required to make fewer decisions. For the 60 mm square HGMF the factor is about 22.

In addition to reducing the number of decisions, the HGMF may also increase the accuracy with which each one is made by improving the signal to noise ratio. This is because the lateral growth restraint tends to make colonies grow taller than normal, thereby enhancing any optical property on which detection is based. There is also evidence that one of the major sources of noise — food debris — may be selectively removed from the HGMF (Sharpe and Michaud, results to be published). This would also improve the signal to noise ratio and the potential accuracy of the counter. Thus, three related factors can operate to ultimately raise the accuracy of the HGMF counting system over that of the conventional random plating or filter system.

A fourth factor may also be particularly valuable in relation to the complete automation of microbiological analyses. This is a requirement for the capability of distinguishing between a very small number of colonies and a very large number (confluent growth in the conventional system). Since the counter for the HGMF need not rely on detecting the boundaries between colony and substrate, any level of microorganisms above that required to saturate the HGMF will always yield the maximum number of positive grid-cells and induce the corresponding maximum count in the counter. Obviously, the counter may be constructed so that any recorded value close to the saturation value activates a warning that the count should be treated with suspicion. Thus the need for human technicians to inspect all plates can be reduced or eliminated.

Isolation of Microorganisms

Devices which will automatically isolate pure cultures of microorganisms are potentially useful in hospitals, typing laboratories, and other locations. For such a device to depend on recognition of colonies, as in a conventional plating system, it must be capable of distinguishing whether a colony is sufficiently far from others and flawless in appearance for it to stand a good chance of being "pure." However, the mechanics and the probabilities associated with conventional plating and isolating techniques are only vaguely definable; the degree of electronic sophistication required (and therefore the cost) to be able to recognize "probably-pure" colonies is relatively high.

The HGMF can be used to advantage as a subcomponent in an apparatus that will scan and mechanically subculture growths from its various grid-cells. Indeed, it may also be used to advantage in manual methods of isolation, since the probability that the growth in any positive grid-cell is pure, or not pure, can be calculated precisely, without the need to determine the position of the growth relative to others, or the flawlessness of its appearance. All positive grid-cells on the HGMF have the same probability of being pure cultures; if x of the grid-cells on an HGMF of N grid-cells are positive, this probability (P)* may be calculated from:

$$P = \frac{(N-x)}{x} \ log_e \ \frac{N}{(N-x)}$$

A graph of "Probability of purity" against number of positive grid-cells for a 10,000 grid-cell HGMF is shown in Figure 10-5. It can be seen that when 200 grid-cells are positive, each growth has a 99.0 percent probability of being pure. When 970 grid-cells are positive, each growth has a 95.0 percent probability of

---

*Actually the maximum likelihood estimate of the probability.

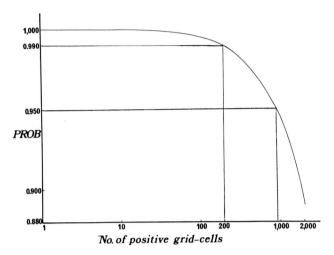

Figure 10-5. Graph illustrating the probability that the growth in any positive grid-cell of a 10,000 grid-cell HGMF is pure, i.e. was derived from only one growth unit. The numbers of positive grid-cells for 95 and 99 percent confidence levels are indicated.

being pure. Thus, using the HGMF, isolation can be carried out to any required degree of confidence simply by adjusting the concentration of organisms, or alternatively, the degree of confidence may be stated after reference to the number of positive grid-cells. The counter/detector described above is obviously suited to controlling or monitoring the rectilinear movements of a subculturing apparatus for the HGMF, or of deciding whether or not subculturing is worthwhile on a given filter.

CONCLUSION

Through its potential for simplifying the engineering problems of mechanizing microbiological counting and isolation processes, by removing the need to make dilutions or improving the reliability of detecting colonies, the HGMF appears to be an attractive subcomponent of mechanized systems. The most useful format would seem to be a square, preferably attached to a rigid and stackable plastic frame which would serve to carry the HGMF into any machine and register it accurately in an electronic counter. Until such a product becomes commercially available, many of the potential uses of HGMFs may be investigated in manual techniques using 47 mm printed circles; the authors will gladly assist interested colleagues by printing moderate quantities of HGMFs for serious studies.

REFERENCES

American Public Health Association: 1971. *Standard Methods for the Examination of Water and Waste Water*, 13th ed., p. 678. American Public Health Association Inc., New York.

Sharpe, A.N. and G.L. Michaud: 1975. Enumeration of high numbers of bacteria using hydrophobic grid-membrane filters. *Appl Microbiol, 30:* 519.

Sharpe, A.N.: 1975. Machine for printing hydrophobic grids on membrane filters. *Appl Microbiol, 30:*110.

Chapter 11

USE OF SPIRAL BACTERIAL PLATING AND LASER COLONY COUNTING TECHNIQUES IN STUDIES OF THE MICROBIAL ECOLOGY OF MAN

W. W. BRINER, J. A. WUNDER, D. W. BLAIR,

J. J. PARRAN, T. L. BLANEY, AND W. E. JORDAN

Abstract

The objectives of this study were twofold: (1) to determine the suitability of spiral plating and laser colony counting devices for rapid large-scale microbiologic sampling and (2) to gather information on the microbial ecology of the skin of infants and the scalp of adults. The spiral plating device and the laser counter appeared to have significant advantages in saving of time, equipment, and technical manpower over the spread plate conventional method. The methods reported were easily adaptable to the handling of large numbers of samples in a microbiological study. The data revealed that the aerobic microbial flora of the infants sampled was primarily coccal in nature, most probably staphylococcal. Staphylococci also played an important role in the microbial ecology of the scalp, as did aerobes and facultatives. Streptococci and gram-negatives did not appear to play a significant role.

INTRODUCTION

WHEN a quantitative determination of various microorganisms in a sample is desired, the samples are usually plated by hand on the agar medium of choice. Standard pour plates or spread plates are the procedures most commonly employed for this. Both procedures demand extensive time, manpower, and equipment. These methods involve serial dilutions of each sample as well as replicate plating of each dilution. After incubation, those plates that have a total colony count between

30 and 300 are manually counted.

The time and equipment required pose enormous logistic problems when a large number of bacterial samples must be plated and enumerated rapidly. In the authors' laboratories, under certain conditions, the conventional methods of plating and counting proved totally inadequate to handle a large volume of samples within reasonable time and cost. Thus, the need for a more efficient manner of delivering bacteria onto the surface of a plate and for enumeration of colonies after incubation was clearly defined.

In 1973, Gilchrist et al. and Campbell and Gilchrist described the development of a semiautomatic rapid (spiral plate) method for delivering a liquid sample containing bacteria onto the surface of an agar plate (see also U.S. Patent No. 3,799,844). The spiral plate method was found by these workers to compare favorably with the pour plate method. A laser-actuated bacterial colony counter has recently been developed (Exotech Corp., Gaitherburg, Maryland) which will rapidly enumerate the colonies on the spiral plate. In these studies the authors used the spiral plating and laser counting techniques to study the microbial ecology of infant human skin and of the human scalp.

MATERIALS AND METHODS
Spiral Plating Machine

The plating method involves depositing approximately 35 μl of sample, in the form of an Archimedian spiral, onto the surface of an agar plate of 150 mm diameter rotating at a constant speed. The sample is delivered at a logarithmically decreasing rate from a point near the center of the plate to a point near the outer edge. A cam, which actuates a 100 μl syringe, controls the delivery rate of sample as the agar plate is turned at a constant rate. Examination of a plate after incubation reveals growth of the bacterial colonies along the line of the spiral. The number of bacteria in the deposited sample determines the concentration of bacterial colonies along the line of the spiral. As the volume of sample delivered per unit length of the spiral is decreased, the number of colonies per unit length of spiral is directly related to that decrease in volume. The incidence of colonies decreases about ten-thousandfold along the course of the spiral track. Therefore, at some point across that colony gradient distinct colonies can be enumerated.

In this study the samples were introduced into the tubing of the spiral plater and dispensed onto agar plates as described by Gilchrist et al. (1973) and by Campbell and Gilchrist (1973).

Laser Counter

Gilchrist and Campbell (1973) developed a counting grid for relating the number of colonies on a spiral plate to the number of cells per unit volume of sample, but the procedure proved quite time consuming. This was overcome by use of a laser-type bacterial colony counter. The counter is based on the principle of projecting a laser beam through a plate of bacteriological agar onto a photodiode tube. The change in the transparency of the plate, caused by the bacterial colonies, registers on the photodiode and activates a counting circuit. The laser beam scans the plate surface in a reverse Archimedean spiral; that is, from the periphery to the center of the plate.

The laser counter reads out in terms of both colonies counted and area of the plate surface counted. Beam size and sensitivity can be selected. These parameters can be adjusted to enumerate colonies of varying diameters. Beam size is the diameter of the laser beam. Sensitivity is a measure of the percentage of the laser beam that a bacterial colony must impede in order to activate the counting circuit. The laser parameters of count and area are inversely related. When one counts a predetermined number of colonies, the counter reads a smaller area of a colonized plate when the colony concentration is heavy and a larger area when the colony concentration is small.

In counting, the usual procedure is to set the instrument to count a certain number of colonies, e.g. 200, and simultaneously the area is read in which that number of colonies was counted. However, the entire plate may be counted when the total number of colonies is low (less than 200). In this experiment 200 colonies were counted at beam size of 200 and sensitivity of 6 and areas were recorded.

Standardization of Spiral Plater and Laser Counter

Standardization curves were made using each plating machine to correlate the area readouts of the counter to the number of bacteria per ml of sample. Concentrations of 10^4 to 10^7 per ml of *Staphylococcus aureus* and *Escherichia coli* were plated by both the spread (Meynell and Meynell, 1965) and spiral plate methods. The spiral plates were counted using the laser counter and the areas were recorded. The spread plates were counted both by the laser counter and manually. The area data from the spiral plate counts were plotted (computer) by the method of least squares against the number of bacteria per ml of culture (*S. aureus* and *E. coli*) as determined by the spread plate method. An equation was written from the resulting curves for each of two machines relating the laser area to the number of bacteria per ml. A computer program was written which would calculate the number of bacteria per ml from the laser area *(see* RESULTS).

Experimental Designs

The objectives of these experiments were twofold:
1. to determine the suitability of the spiral plating and laser counting methods for rapid large-scale microbiological sampling.
2. to gather information on the microbial ecology of the skin of infants and on the microbial ecology of the scalp of adults.

Two experiments were performed. Each served both objectives.

Experiment 1: Study of the Microbial Ecology of the Skin of Infants in the Inguinal Area

SUBJECTS: The subjects consisted of 128 infants three to eighteen months of age and of 12 to 24 lb body weight. Sixty-four of the infants were sampled on each of two consecutive days.

SAMPLING AND PLATING: One square inch of skin in the inguinal region of each infant was sampled by swabbing for 1 min with a Dacron® swab (Scientific Products Company, Columbus, Ohio) saturated with 0.1 percent Triton X-100® buffered to pH 7 with 0.075M phosphate buffer (Williamson and Klig-man, 1965). The swab was agitated for 30 sec in 4.0 ml of Triton/ phosphate buffer using a Vortex Mixer® (Scientific Industries, Queens Village, New York) and the resulting suspension was plated onto the various media using the spiral plating machine.

MEDIA: The media (BBL) used in this experiment are pre-sented in Table 11-I. These were poured by hand, inoculated

TABLE 11-I

MEDIA USED IN EXPERIMENT 1

Medium	*Abbreviation*	*Selective for:*
Phenylethyl Alcohol Agar	PEA	Micrococci and streptococci
Mannitol Salt Agar	MS	Staphylococci
Tergitol Agar	T7	Gram-negative organisms
Tripticase Soy Agar	TSA	Nonselective

using the plater, and incubated aerobically 18 to 24 h at 37°C. Plates were counted with the laser counter, except for Tergitol 7® plates which were enumerated by hand because of the small number of colonies. Two plates of each medium were inocu-lated.

PARAMETERS ESTIMATED: To accomplish objective 1, the authors assessed the effects of different spiral plating machines (N = 2) and operators (n = 4, two experienced and two inex-perienced in the operation of the spiral plater) on microbial counts. The study was balanced by operator and machine. The sample (from one subject) was assigned at random to one of two plating machines and the operators were assigned at random to each of the plating machines. Estimates of the time required to plate samples, amount of medium used, and glassware needed were made in order to compare the spread plate and the spiral plating methods. Objective 2 was accomplished by assessment of the microbial data.

STATISTICAL ANALYSIS: After the data were calculated as number of bacteria per ml, a logarithmic transformation was made in order to achieve a more nearly normal distribution. The data were then analyzed by an analysis of variance (computer) and the variables (machines = 2, operators = 4, and media = 3) were ranked by the method of Newman and Keuls as described by Snedecor and Cochran (1967) in order that all comparisons among means could be made while maintaining a constant risk of $\alpha = 0.05$.

Experiment 2: Study of the Microbial Ecology of the Scalp

Thirty adult, male coworkers volunteered for this study.

SAMPLING AND PLATING: Each subject was sampled by placing 20 ml of Tween 20® on the scalp during a vigorous 30 sec massage by a trained beautician. Sterile distilled water (900 ml) was poured over the scalp slowly while the scalp was being massaged vigorously. The process was repeated. This rinse water was collected in a sterile plastic vessel, pooled, and an aliquot was withdrawn for plating.

The spiral plating was done as described above. For spread plates, dilutions were made depending upon the medium (Table 11-II) as follows: trypticase soy agar (TSA), 10^{-1} to 10^{-6}; Vogel-Johnson (V-J), none to 10^{-1}; Schaedler's (SCH), 10^{-2} to 10^{-5}. Each dilution, 0.2 ml, was spread evenly upon the surface of a single plate of the appropriate medium. The sample was delivered with a 1.0 ml plastic pipet (BBL).

MEDIA: The media (BBL) used in this experiment, and their abbreviations, are listed in Table 11-II. The media were poured using the Petrimat® (Streuers, Copenhagen, Denmark) automatic media dispensing apparatus, which pours up to 600 plates per h. The SCH medium was poured 48 h before use and stored under anaerobic conditions using the Gas Pak (BBL) method. Anaerobic incubation of inoculated plates was under the same conditions. Incubation conditions were as listed in Table 11-II. Spiral plates were counted using the laser, spread plates by hand.

PARAMETERS ESTIMATED: To accomplish objective 1, the spiral plate method was compared with the spread plate method using three different media, TSA, V-J, and SCH. Various media were also compared with each other for their efficiency of cultivation of microorganisms from these samples. The comparisons, grouped together for convenience in Table 11-II were total

TABLE 11-II.

MEDIA COMPARED FOR PLATING EFFICACY

Medium	Abbreviation	Selective for:	Purpose	Incubation at 37°C Conditions	Time
Trypticase Soy Agar	TSA	Gram-negatives	Total aerobes	Aerobic	24-48 h
Trypticase Soy Agar +		Nonselective			
0.1% Tween 80	TSA + Tween		Total aerobes	Aerobic	24-48 h
Vogel-Johnson Agar	V-J	Nonselective	Staphylococci	Aerobic	24-48 h
Staph 110 Agar	S110	Staphylococci	Staphylococci	Aerobic	24-48 h
Tergitol 7 Agar	T7	Staphylococci	Gram-negatives	Aerobic	18-24 h
TSA + 0.1% Na					
Alkyl Sulfate	TSA + SAS	Gram-negatives	Gram-negatives	Aerobic	18-24 h
Eosin Methylene					
Blue Agar	EMB	Gram-negatives	Gram-negatives	Aerobic	24-48 h
Streptosel Agar	Strep	Streptococci	Streptococci	Aerobic	24-48 h
H.L. Ritz Agar*	HLR	Streptococci	Streptococci	Aerobic	24-48 h
Schaedler's Agar	SCH	Nonselective	Total anaerobes	Anaerobic	7 days

*Ritz (1967)

anaerobes, TSA versus TSA plus Tween 80; staphylococci, V-J versus S110; gram-negatives, T7 versus TSA plus SAS and versus EMB; streptococci, HLR versus strep. Objective 2 was accomplished by assessment of the microbiological data.

STATISTICAL EVALUATION: After the data were calculated as the number of bacteria per ml of sample fluid, a logarithmic transformation of the data was made in order to achieve a more nearly normal distribution. An analysis of variance was performed on the transformed data comparing methods (n = 2, spread and spiral) across media (n = 3, TSA, V-J and SCH). Data on the gram-negatives and streptococci were not treated statistically because the large number of zero counts precluded evaluation of these data.

RESULTS

Standardization Curves of Spiral Plating Machine and Laser Counter

Standardization curves were established for each plating machine using a gram-positive microorganism *(Staphylococcus aureus)* and a gram-negative microorganism *(Escherichia coli).*

Each curve related the laser area to the number of bacteria per ml of culture, as determined by the spread plate method. Theoretically, there should be no significant difference between the curves obtained from the two microorganisms on the same plating machine over a similar range of bacterial concentrations (in this case 10^4 to 10^7 per ml). The results showed this to be the case; therefore, the data were pooled to produce a master curve for each spiral plating machine as shown in Figures 11-1 and 11-2. The general equation for the curve is:

$$\log_{10} \text{bacteria/ml} = a + b \text{ (laser area (} + c \text{ (laser area)}^2$$

where a, b, and c are derived constants, specific for each plating machine. A computer program which converted laser area readings into number of bacteria per ml was written for each curve.

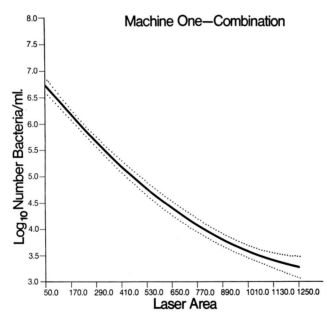

Figure 11-1. Calibration curve for spiral plating machine number 1. The $\log_{10}$ number of bacteria/ml of sample plotted against the laser area using the method of least squares.

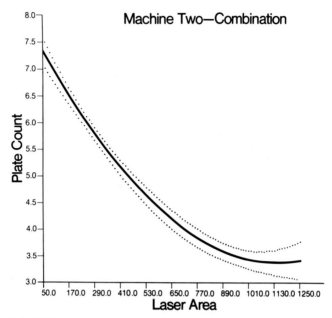

Figure 11-2. Calibration curve for spiral plating machine number 2. The $\log_{10}$ number of bacteria/ml of sample plotted against the laser area using the method of least squares.

Experiment 1: Microbial Ecology of Infant Skin

COMPARISON BETWEEN THE TWO SPIRAL PLATING MACHINES: There was no significant difference between machines (Table 11-III) in this study.

TABLE 11-III

EXPERIMENT 1

COMPARISON BETWEEN SPIRAL PLATING MACHINES

(Across Subjects = 128, Operators = 4, and media = 3)

Plating Machine Number	Mean Log No. of Bacteria/ml of Sample Fluid
1	5.2706*
2	5.2443

*Variables within brackets are not significantly different from one another at $\alpha = 0.05$, least significant difference = 0.8879 units.

OPERATORS: Comparisons were also made among the mean log bacteria per ml counts of four plating machine operators (two experienced and two inexperienced). There was no significant difference observed among the operators (Table 11-IV).

TABLE 11-IV

EXPERIMENT 1

COMPARISON AMONG OPERATORS

(Across Subjects = 128, Media = 3, and Machines = 2)

Operator Number	Mean Log No. of Bacteria/ml of Sample Fluid
1 (Experienced)	4.9787*
4 (Inexperienced)	5.0992
2 (Inexperienced)	5.2473
3 (Experienced)	5.7043

*Variables within brackets are not significantly different from one another at $\alpha = 0.05$, least significant differences = 1.2557 units.

MICROBIAL ECOLOGY OF SKIN: Comparisons were also made among the counts obtained from three of the four media tested (Table 11-V). Those compared were MS, TSA, and PEA. There was no significant difference among counts from each of the three media above ($\alpha = 0.05$), although counts from MS were slightly lower numerically than those of TSA and PEA.

TABLE 11-V

EXPERIMENT 1

COMPARISON AMONG MEDIA

(Across Subjects = 128, Operators = 4, and Machines = 2)

Medium	Microorganisms Assayed	Mean Log No. of Bacteria/ml of Sample Fluid
Mannitol Salt Agar	Staphylococci	4.9810*
Trypticase Soy Agar	Total aerobes	5.3692
Phenylethyl Alcohol Agar	Staphylococci and streptococci	5.4219

*Variables within brackets are not significantly different from one another at $\alpha = 0.05$, least significant difference = 0.5274 units.

In most cases, the number of gram-negative colonies found on T7 medium was very low. There were some samples, however, which would have yielded a significant count, but the colonies grew confluently. These data were considered unreliable and were omitted from the above statistical analyses.

RAPIDITY OF OPERATION OF SPIRAL PLATING MACHINES AND LASER COUNTER: It was observed that one operator, using two machines, could inoculate about 100 plates per h. The laser counting procedure and computation of number of bacteria per ml required about 2 min per sample (six plates were counted on each sample). (The Tergitol 7 plates were not counted on the laser because of the small number of colonies). One day was required to count and calculate the data.

Experiment 2: Microbial Ecology of the Scalp

SPREAD VERSUS SPIRAL PLATING METHODS: The comparison between methods and the contributions of each medium to the effect is shown in Table 11-VI. Significantly higher counts ($\alpha = 0.05$) were obtained with the spiral plating method. TSA showed no difference between methods, but V-J and SCH showed

TABLE 11-VI

EXPERIMENT 2

COMPARISON BETWEEN SPIRAL — AND SPREAD-PLATE METHODS

Method	Medium	$\overline{X}$ Log No. Bacteria/ml
Spiral	SCH	5.93*
Spread	SCH	5.73
Spread	TSA	4.66
Spiral	TSA	4.59
Spiral	V-J	3.17
Spread	V-J	2.58
Least significant difference ($\alpha = 0.05$) = 0.15 units.		
Spiral All 3 (SCH, TSA, V-J) Combined		4.57
Spread All 3 (SCH, TSA, V-J) Combined		4.32
Least significant difference ($\alpha = 0.05$) = 0.10 units.		

*Treatments within brackets are significantly different from those outside at $\alpha = 0.05$.

significantly higher counts with the spiral method. However, both methods showed SCH > TSA > V-J in counts per ml of sample fluid.

COMPARISON OF MEDIA FOR PLATING EFFICIENCY: The data are presented in Table 11-VII. TSA produced counts significantly higher ($\alpha = 0.05$) than TSA plus Tween. S110 produced counts significantly higher ($\alpha = 0.05$) than V-J. Gram stains of

TABLE 11-VII

EXPERIMENT 2

COMPARISON OF PLATING EFFICIENCIES BETWEEN MEDIA

Total Aerobic Media		*Staphylococcal Media*	
Medium	$\overline{X}$ *Log No. Bacteria/ml*	*Medium*	$\overline{X}$ *Log No. Bacteria/ml*
TSA	5.93*	S110	4.40
TSA + Tween	5.06	V-J	3.20

*Treatments within brackets are significantly different from those outside at $\alpha = 0.05$, least significant difference ($\alpha = 0.15$) units.

representative colonies from S110 and V-J showed that S110 permitted growth of cocci arranged in tetrads, which were not found on V-J. The media for gram-negatives (T7, TSA plus SAS, and EMB) and streptocococi (Strep and HLR) yielded many plates with zero or less than thirty colonies, which precluded valid statistical analysis of these data.

MICROBIAL ECOLOGY OF THE SCALP: The numbers of gram-negative organisms and streptococci were too low to permit valid statistical analyses. With both the spread plate or spiral plate method there were higher counts on SCH > TSA > V-J.

DISCUSSION
Suitability of Spiral Plater and Laser Counter for Large-Scale Studies

The data indicated that the spiral plating method, used in conjunction with the laser colony counter, has significant advantages over the pour plate and the spread plate methods for enumerating bacteria. These advantages include savings of time, manpower, and materials.

Saving of time is a most important advantage of this system. It took approximately 1 min to dispense one plate. In the skin study, sixty-four samples (eight plates per sample) were studied in a period of about 6 h. Perhaps the greatest time saving occurred with the use of the laser colony counter. Only 2 to 3 sec were required to scan a plate and obtain the laser area readout. Laser areas were recorded on paper and later entered into a computer for conversion to numbers of bacteria per ml. The entire counting process and computation required less than 2 min per sample. If the conventional methods for plating and counting had been used the authors estimate the time would have been about seven– to tenfold greater per sample. Thus, for 128 subjects (one sample per subject), only two days were required for plating and one day for counting and calculations.

There are savings in materials using the described methods. In the scalp study using the spread plate method, one plate per dilution was made on TSA dilutions of 10^{-1} to 10^{-6} of the scalp washings for a total of six plates per sample; this required 156 plates for twenty-six subjects. Of these 156 plates, 130 were discarded for having too few (<30) or too many (>300) colonies and twenty-six were counted. Using the spread plate method 83 percent of the plates were discarded without contributing to the data. Assuming that about 20 ml of medium were used for each plate, for the spread plates discarded 2.6 l of media were consumed for no data output. The thirty plates inoculated and counted by a spread plate method represent 0.6 l of media consumed. The twenty-six plates used for the spiral plating machine (45 ml per plate) required 1.2 l of media. Thus the spiral plating method (1.2 l) required 61 percent less media than the spread plate method (3.2 l). The spiral plate method also required 83 percent less plates (26) than the spread plate method (156). Also, 156 dilution tubes were required using the spread plate method while no dilution tubes were needed using the methods in this experiment. There was also no need for twenty-six pipets and twenty-six bent glass rods (for spreading plates) using the spiral plates. No special water bottles, dilution bottles, etc. were required. (Comparative saving would grow less when fewer dilutions were plated using the spread plate method). The spiral plating machines are portable and can be transported with ease to a study site.

Another significant advantage of these rapid methods is the potential saving in manpower. The inexperienced machine operators used in this study were unfamiliar with bacteriological techniques and had not seen a plating machine prior to the study. Each of these operators was given only 15 min of instruction before the study began, yet there was no significant difference between the mean bacterial counts obtained by these inexperienced operators and those obtained by two experienced operators. These data support those of Gilchrist et al. (1973). Also, one person can easily operate two plating machines simultaneously. The data from the skin study (Table 11-III) showed no significant difference between machines, with respect to enumeration of the microbial populations examined.

The direct comparison between the spread and spiral plating methods (Table 11-VI) revealed that the spiral plating method produced higher counts than the spread plate method. A similar finding was observed comparing the pour plate method with the spiral plate method of Gilchrist et al. It is of interest to note that Clark (1967) showed that the spread plate method gave counts 70 to 80 percent higher than the pour plate method on samples of water, ice, and processed birds (from poultry plants). One reason proposed for the higher counts by the spread plate method was the breaking up of chains and clumps during spreading. It is possible that in the spiral plating method, even further breaking up of these chains and clumps occurred during the passage of the sample through the orifice of the stylus and its deposition onto the agar. The data in this paper, and those of Gilchrist et al. and Clark, suggest that the magnitude of difference between methods may be a function of the kind of organisms sampled and the media used. This point bears further investigation.

Comparison of Media (Experiment 2)

The data in Table 11-VII indicate that TSA was superior to TSA plus Tween and S110 to V-J with respect to counts per ml of sample fluid. The addition of Tween to TSA resulted in an inhibition of growth of some of the microorganisms from the scalp. Since TSA is a nonselective medium, addition of Tween certainly is not an advantage since it increases selectivity. Similarly, V-J appeared more inhibitory than S110 to certain microorganisms from the scalp. (These appeared to be cocci arranged in tetrads). In a future experiment the choice between S110 and V-J would depend upon whether or not the investigator wants a more or less selective medium.

Microbial Ecology of the Skin in Infants in the Inguinal Area (Experiment 1)

Mannitol salt agar is selective for staphylococcus (Chapman et al., 1938) and PEA will support the growth of micrococci and streptococci (BBL Manual, 1973). Since there was no significant difference between the two media with respect to number of organisms per ml of sample fluid, it is suggested that the microflora of these infants was primarily coccal in nature, most probably staphylococcal. This conclusion is also supported by the fact that there was no significant difference between counts on TSA (a nonselective medium), MS, and PEA. Furthermore, a few gram-negatives were found on Tergitol 7. Had other gram-positive or gram-negative aerobes been present in significant numbers, higher counts on TSA and Tergitol 7 should have been observed. However, further studies, including studies on anaerobes, would be necessary to define more precisely the microbial ecosystem on the inguinal skin of infants.

Microbial Ecology of the Scalp (Experiment 2)

With respect to the microbial ecology of the scalp, gram-negatives and streptococci appeared to play little role in this ecosystem by virtue of their low numbers. Staphylococci (V-J medium) appeared to have an important position in the scalp ecosystem, as did facultatives and anaerobes, since the counts on

SCH were significantly greater ($\alpha = 0.05$) than TSA. Obviously, more detailed studies on this ecosystem are necessary if it is to be characterized more completely.

REFERENCES

BBL Manual of Products and Laboratory Procedures, 5th ed. 1973. Cockeysville, Maryland, pp. 132.

Campbell, J.E. and J.E. Gilchrist: 1973. Spiral plating technique for counting bacteria in milk and other foods. In *Developments in Industrial Microbiology,* Vol. 14. American Institute of Biological Sciences, Washington, D.C., pp. 95.

Chapman, G.H., C.W. Lieb, and L. Curcio: 1938. Use of bromthymol blue agar and phenol red mannitol agar for the isolation of pathogenic types of straphylococci. *Am J Clin Path, 8:*3.

Clark, D.S.: 1967. Comparison of pour and surface plate methods for determination of bacterial counts. *Can J Microbiol, 13:*1409.

Gilchrist, J.E., J.E. Campbell, C.B. Donnelly, J.T. Peeler, and J.M. Delaney: 1973. Spiral plate method for bacterial determination. *Appl Microbiol, 25:*244.

Meynell, G.G. and E. Meynell: 1965. *Theory and Practice in Experimental Bacteriology.* Cambridge University Press, London, pp. 15.

Ritz, H.L.: 1967. Microbial Population Shifts in Developing Human Dental Plaque. *Arch Oral Biol, 12:*1561.

Snedecor, G.W. and W.G. Cochran: 1967. *Statistical Methods.* Iowa State Pr, Ames, pp. 273.

Williamson, P. and A.M. Kligman: 1965. A new method for the quantitative investigation of cutaneous bacteria. *J Invest Derm, 45:*198.

Chapter 12

A SEMIAUTOMATIC INSTRUMENT FOR THE DETERMINATION OF GEL RIGIDITY IN MICROBIOLOGICAL NUTRIENTS AND GELLING AGENTS*

I. D. COSTIN

Abstract

The paper describes an instrument designed to measure simultaneously the rigidity and elasticity of gels (usually of an agar and gelatin base) for control laboratories, manufacturers of gelling substances and nutrients, the food industry, as well as microbiological and other laboratories.

INTRODUCTION
Significance of the Determination of Gel Strength (Rigidity)

AGAR and gelatin share the property of ability to form stable gels in dilute solutions. Apart from their use in microbiology, these gels are also used widely by the food, drug, and chemical industries. The most important quality of an agar and gelatin gel for the microbiologist is its consistency or bearing capacity (gel stability, gel strength). Elasticity also plays a role in some applications. Consistency is the major criterion in evaluating the quality and suitability of agar and gelatin for microbiological purposes. Apart from this major feature, physical, chemical, and biological characteristics, such as purity; granulation; solubility; clarity of solutions; pH; liquefaction temperature; gelling temperature; gelling time; Ca^{++}, Mg^{++}, and $Fe^{++/+}$ content; presence of antibacterial substances, peroxides,

---

*The author is indebted to Mr. B. Lippold for his conscientious collaboration in these experiments.

fatty acids, colloidal sulfur, heavy metals, etc.; number and type of microorganisms present, are also taken into consideration. There is need for the capability to determine the gel solidity of these substances accurately especially with the constantly growing importance of tests and the tendency towards development of more uniform and standardized assays for microbiological tests. The factor of gel strength in solid nutrients must not only be taken into consideration by manufacturers, but also by the users of these products. Presumably gel rigidity not only plays a role in regard to inoculation, but probably also in some other processes vital in the culturing of microorganisms, e.g. the size of colonies and their appearance, swarming capacity, diffusion of bacterial metabolites and active substance, determination of resistance, etc. In this context it is surprising to note how little the factor of gel rigidity has been taken into consideration in diagnostic and microbiological research. While the actual agar concentration is almost always indicated, true gel strength, which need not necessarily parallel agar concentration, is rarely mentioned at all. The influence of gel rigidity on the growth of bacterial colonies has been described by Fulthorpe (1951), who found that gel consistency exercises an essential influence on size, form, surface, smooth/rough proportion, and agglutinability. On the other hand, Frank and Buchmayer (1961) concluded that gel consistency has no influence whatever on the number of colonies. Although the influence of gel strength on the results of microbiological antibiotic determination and clinical resistance tests is well known, this factor has also been neglected in the best known publications (Grove and Randall, 1955; Brock, 1957; Bechtle and Scherr, 1958; Bauer et al., 1966; Yousef et al., 1967; Hanus et al., 1967; Ericsson and Sherris, 1971; Barry and Fay, 1973). Conceivably the absence of a handy, adequate instrument for the measurement of gel consistency in microbiological laboratories may explain this omission. Even manufacturers of ready-made microbiological nutrients give no data on gel consistency for agar containing products.

Methods for Testing Gel Strength
(Gel Consistency, Gel Stability)

Over a period of time, numerous methods and instruments have been proposed for determining gel consistency, primarily in industry. Without making any attempt to give an exhaustive presentation, some of these methods are described below.

Sheppard et al. (1920) used a rotating dynamometer and expressed gel rigidity as the breaking load. Lockwood and Hayes (1931) proposed a method based on the degree of deformation (shortening) of an agar cylinder upon removal from a beaker. What is measured in this case is the internal cohesion of the gel mass expressed in percent of the shortening of an agar column. The instrument proposed by these authors was called a "ridge-limeter." Kizevetter (1937) used a method depending on the breaking of a gel by the gradually increasing weight applied to its surface as mercury was added drop by drop to a vessel. The weight was measured through a rod with its bottom resting on the gel surface.

British Standard Specification (1944) proposed the so-called "Bloom Gelometer" as a standard method. According to this method the rigidity of gels is expressed by the weight in grams that is required to produce an indentation of 4 mm with a metal rod, 12.7 mm in diameter, at $+10°C$ and a gel concentration of 6.66 percent.

Campbell (1938) used the principle of gel deformation where a metal rod is sunk into a cylinder of gel prior to gelling; after the gel has congealed an increasing torque is applied to the rod by a system of pulleys. What is measured is the rigidity modulus expressed in CGS units. Drew (1946) proposed a method closely resembling that of Campbell, with the difference that the results were expressed in g per cm^2.

Chakraborty (1948) sank the pan of a balance in a suitable vessel, filled the vessel with agar which was allowed to set, then observed the weight required to tear the pan out through the gel. Heicken (1948) measured the breaking weight of starch gels poured into dishes loaded with a pestle of 0.58 cm^2 basal surface during the increasing addition of weight. In principle, this corresponds to Kizevetter's approach.

Fulthorpe (1951) judged gel rigidity by the weight required by a glass quadrangle with a basal surface of 10 × 10 mm to break through the agar surface within 12 sec, the so-called "12 second value."

Jones (1956) used an indirect method based on determination of the agar concentration necessary to produce an agar jelly of a given strength when prepared under standard conditions. The standard jelly strength selected is one of 75 g, for a deflection of 20 degrees on the British Food Industries Research Association's jelly tester.

Goring (1956) described a so-called "semimicrogelometer" for the determination of breaking resistance and the rigidity of gels. A motor driven tactile scanner of variable speed and diameter was pressed against a gel cylinder placed on a weighing pan. The breaking resistance was given by the load at which the tactile device tore through the gel surface.

Literature on other methods and further descriptions of instruments for the measurement of gel rigidity have been given by Goring (1956). To some extent agar manufacturers and dealers use other, more or less improvised, methods; the so-called "Japanese method" and "Moroccan method" are based on the techniques of Kizevetter (1937) or Heicken (1948) with some minor modifications. Ready-made instruments, such as the Nikan® instrument (Kiya Seisakusho Ltd., Bunkyo, Japan) are also available. The Nikan instrument is based on Fulthorpe's principle (1951). All the above mentioned methods and instruments are encumbered by one or several of the following disadvantages, which limit their usefulness and prevent their routine use in microbiological and most other control laboratories:

    (a) evaluation and measurement of the adhesion of other substrates rather than of the internal molecular cohesion of gels;

    (b) complicated instruments and time-consuming measuring procedures;

    (c) time-consuming calculations;

    (d) impossibility of measuring rigidity and elasticity (deformability) simultaneously.

The instrument of Goring (1956) would seem to be the most favorable one for control laboratories of agar manufacturers, food and drug companies, media manufactures, etc. However, in the author's opinion, it also has defects (such as gel containers which are too narrow, thus affecting the results to a great extent), complicated calculations, and the impossibility of preparing a load/time curve.

The Gelometer

The aim was to develop an instrument which would facilitate a rapid and reproducible measurement of the rigidity and elasticity of gels, especially agar and gelatin-containing substrates. It was intended that the instrument should serve for measuring the rigidity of bacteriological nutrients as well as determine gelling power and elastic properties of the raw materials: agar-agar, gelatins, and possibly other gelling agents. The version selected at the end of numerous experiments was built by an instrument company in the form of a compact laboratory instrument. The instrument is available from the Heinrich Bareiss Apparatebau Company, 7931 Oberdischingen, under the name Gelomat®.

Principle

Since terms like consistency, bearing capacity, stability, rigidity (solidity), deformability, and elasticity are hard to define in exact units in connection with gels, an attempt was made in the elaboration of this instrument to determine two elements: the consistency, bearing capacity, or rigidity, i.e. the mass in g required at constantly increasing load until a probing device with a certain surface area and shape of surface tears through the gel; and the elasticity, i.e. the resistance of the gel against the same constant, progressive load, expressed by a curve in which the load time/pressure ratio is illustrated by comparison with the standard curve of a completely incompressible body. With this instrument both parameters can be determined simultaneously in one continuous reading.

The instrument pushes a test table or plate (on which the test object is placed) vertically upward at a constant rate, and a pressure measuring device (pressure foot-column gauge) measures the load on the column at the particular instant at which the gel surface breaks.

MATERIALS AND METHODS

Construction

The instrument consists of a pedestal to which a rectangular test table with a revolving spindle is attached. The instrument is equipped with a carrying column, the loading device, and a special dial gauge with a trailing pointer. The measuring scale ranges from 0 to 200 g. Each division on the scale corresponds to 2 g. Further details of the construction can be obtained from Figures 12-1 and 12-2.

Results are read off in seconds. The table feed is 0.475 mm per 5 sec; thus 1 sec is equivalent to a penetration depth of 0.095 mm. To test agar for microbiological purposes, the instrument is fitted with a cylindrical probe with a bottom surface of 10 mm^2. For substances having gel stabilities below 40 or above 100 g, tactile probe of other dimensions may be used. A leveling indicator (clinometer) and four adjustable supports allow the instrument to be set up horizontally. The instrument operates on 220 V alternating current and weighs approximately 10 kg (Fig. 12-2).

Working Procedure

Numerous preliminary experiments showed that the measured gel stability of agar or gelatins may be affected by certain factors. A standardized measuring technique requires adequate consideration of all these factors:

(a) concentration of gelling substance
(b) composition of solvent (water, electrolytes, other solutions)
(c) pH of solvent
(d) technique of preparation (method of dissolution, boiling time, autoclave temperature and time, pouring tempera-

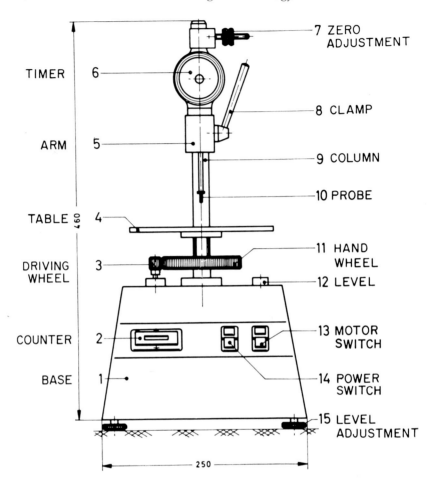

TIMER 6

ARM 5

TABLE 4

DRIVING WHEEL 3

COUNTER 2

BASE 1

7 ZERO ADJUSTMENT

8 CLAMP

9 COLUMN

10 PROBE

11 HAND WHEEL

12 LEVEL

13 MOTOR SWITCH

14 POWER SWITCH

15 LEVEL ADJUSTMENT

Figure 12-1. Diagram of the Gelomat instrument.

ture, cooling temperature and time)

(e) height of gel layer

(f) condition of gel surface (dry, damp, or wet)

(g) measuring technique (temperature of gel and of air measuring site, continuous or interrupted load, shape and size of tactile probe).

Not all these factors are affected uniformly; the agar concentration, pH of the gel, condition of the gel surface, and measuring

Figure 12-2. The Gelomat instrument.

technique are extremely important factors. The so-called standard measuring techniques to assess the gel stability of agar for
microbiological laboratories is illustrated below. With some
minor modifications, the same method may also be applied to
gelatins.

Gel Preparation

Agar (12 g of shreds or powder) is added to 1000 ml of cold
Nutrient Broth (Merck, type #7882), allowed to swell for 15
min and subsequently heated for about 45 min in a vapor bath,
with occasional shaking to dissolve the agar completely. The
agar should be fully dissolved; no "schlieren" should be visible
in the liquid, nor should any agar particles adhere to the wall of
the flask when it is shaken. After heating, the flask is autoclaved

for 15 min at 121°C, and cooled in a water bath to 60°C. The pH is determined electrometrically and adjusted to 7.2 if necessary with 1 N NaOH or 1 N HCl. Sufficient agar solution is poured into Petri dishes or other suitable shallow vessels to produce a 1 cm high layer. The vessels are allowed to stand at room temperature in a vibration free area while gelling occurs. The plates are then dried, without cover, for 1 h in an incubator at 37°C and subsequently left for 3 h at room temperature.

Gel Stability Measurements

Prior to measuring, the horizontal adjustment of the instrument is checked and adjusted if necessary with the aid of the binding bolts (15),* and the clinometer (12). The instrument is then adjusted to indicate zero with the zero point set screw (7). A plate of gel is placed on the test table (4), which is elevated by rotating the handwheel until the gel surface just barely touches the bottom of the tactile probe (1). Care must be taken to assure that the trailing pointer of the dial gauge is set in place first, before the measurement indicator needle. The pressure head (14) is then activated and the motor drive and the chronometer switched on (13). The test table begins to move upward with the plate, and the tactile probe is loaded continuously. The load continues and is registered by the two indicators (measuring indicator and trailing pointer) on the dial gauge. As long as the gel surface remains intact, the indicators move simultaneously; at the moment where the test tip penetrates the gel surface the measuring indicator needle returns to point zero, whereas the trailing pointer stops at the penetration value determined. This value, expressed in grams, represents the actual gel stability (bearing capacity, gel rigidity). At the same time, the counter (2) indicates the time elapsed in seconds until the moment of penetration. Once penetration has taken place, the drive and timer are switched off, and the test table together with the sample are retracted using the handwheel. The temperature of the sample is now determined with a piercing thermometer, and if it deviates from 20°C, the standard value for 20°C is de-

---

*Numbers in parenthesis in this section refer to designation in Figure 12-1.

termined with the aid of the correction table (Table 12-I). (Example: gel temperature at time of measurement = 24°C, gel stability measured 38 g; 38 × 1.13 = 42.94 g = actual gel stability). Three different measurements are performed for each sample:

(a) at a distance of 15 mm from the edge of the plate,
(b) at a distance of half the radius from the edge, and
(c) in the center of the plate.

TABLE 12-I

CORRECTION TABLE FOR GEL STABILITY DETERMINATIONS

Gel temperature (room)°C	Correction factor	Gel temperature (room)°C	Correction factor
10	0.74	21	1.03
11	0.76	22	1.06
12	0.79	23	1.09
13	0.81	24	1.13
14	0.84	25	1.16
15	0.86	26	1.20
16	0.89	27	1.23
17	0.91	28	1.27
18	0.94	29	1.31
19	0.97	30	1.35
20	1.00	31	1.39

Measuring points are at least 25 mm from each other. The mean value from these three measurements is calculated, and corrected for temperature. For common microbiological nutrients and gelling substances, a standard cylindrical probe with a contact surface of 10 mm² is used.

The following changes are made when gelatin gels are measured:

(a) the gelatin concentration is about 15 percent,
(b) after pouring, plates are stored in the refrigerator at 6°C for 2 h, and afterwards for approximately 30 min at room temperature,
(c) at the time of measurement, the gelatin gel temperature should be between 15 and 20°C.

Probes with a larger contact surface may be used for softer gels.

In preliminary experiments, tactile probes of various shapes (cylinders, rectangles, spheres) were tested; the cylindrical form proved best for microbiological nutrients and gels.

RESULTS AND DISCUSSION

Sensitivity

The Gelomat instrument has a sensitivity of approximately 2 g; however, depending on whether the load is continuous or interrupted, different results are achieved on the same gel. Higher values are obtained for a continuous load (standard method) than for intermittent loads.

Measuring Limits

With the standard equipment (dial gauge of 0 to 200 g and a cylindrical probe with a contact surface of 10 mm^2), the measuring limit is between 0 and 200 g. The instrument may be used for harder gels provided the dial gauge is replaced by one standardized for heavier loads.

Reproducibility of Results

Provided the conditions described above are strictly maintained, gel stability figures obtained by the Gelomat instrument are largely reproducible. Table 12-II, III, and IV list the results from ten parallel tests made on each of three different agars. In these tests the values for break-through load have not been listed.

Quantitative Performance

For common microbiological nutrients and gelling substances, each measurement takes between 40 and 60 sec. Ordinarily, three measurements at three different spots are done for each sample. One person can conveniently measure fifteen samples per hour allowing for time needed for placing plates on the test table, and introducing and removing the measured samples.

Determination of Gel Elasticity

Apart from the bearing capacity, it is also possible to evaluate the elasticity of a gel with the "Gelomat." For this purpose, the

TABLE 12-II
GEL STABILITY MEASUREMENTS ON AGAR, MERCK TYPE #1613
(STANDARD METHOD)

Time (sec)	5	10	15	20	25	30	35	40	45
Measurement No.	Measured load (g)								
1	4	11	17	25	32	40	45	50	55
2	5	12	19	26	33	42	46	51	56
3	5	12	19	26	32	38	47	52	54
4	5	12	19	26	32	41	45	50	55
5	5	12	18	25	31	40	45	50	55
6	6	12	19	26	32	40	46	51	56
7	5	12	19	26	32	40	46	51	56
8	6	13	20	26	32	41	47	52	54
9	5	11	18	25	32	41	46	51	56
10	5	11	18	25	32	40	47	52	54
Mean value	5.1	11.8	18.6	25.6	32.0	40.3	46.0	51.0	55.1
Standard deviation	0.57	0.63	0.84	0.52	0.47	1.06	0.82	0.82	0.88

TABLE 12-III
GEL STABILITY DETERMINATIONS ON AGAR, MERCK #1617
(STANDARD METHOD)

Time (sec)	5	10	15	20	25	30	35	40	45
Measurement No.	Measured load (g)								
1	4	10	16	22	27	32	37	44	51
2	4	10	16	22	28	33	38	43	52
3	4	10	15	22	27	32	36	43	51
4	4	10	15	21	28	32	36	44	51
5	5	10	17	22	29	33	37	44	50
6	5	10	17	23	29	31	37	42	51
7	4	10	16	22	28	31	37	43	50
8	5	10	17	23	30	32	38	43	51
9	4	10	16	22	30	33	36	44	52
10	5	10	16	22	27	32	37	44	51
Mean value	4.4	10.0	16.1	22.1	28.3	32.1	36.9	43.4	51.0
Standard deviation	0.52	0.00	0.74	0.57	1.16	0.74	0.74	0.70	0.67

TABLE 12-IV

GEL STABILITY DETERMINATIONS ON AGAR, MERCK #1614
(STANDARD METHOD)

Time (sec)	5	10	15	20	25	30	35	40	45
Measurement No.	*Measured load (g)*								
1	5	12	20	27	33	42	46	54	61
2	6	12	20	27	33	41	47	55	60
3	6	14	21	27	34	42	46	54	60
4	6	13	20	27	33	42	46	53	61
5	6	13	20	27	33	42	47	54	62
6	6	13	20	27	34	42	48	54	61
7	6	13	20	27	35	43	47	53	61
8	6	14	21	28	34	43	47	54	60
9	6	13	21	28	34	43	46	50	62
10	6	14	21	28	34	44	48	54	61
Mean value	5.9	13.1	20.4	27.3	33.7	42.4	46.8	53.5	60.9
Standard deviation	0.32	0.74	0.52	0.48	0.67	0.84	0.79	1.35	0.74

load is plotted against time. Figure 12-3 illustrates as example the measurement of agar and gelatin by comparison with an incompressible body. The load, in grams, is entered on the ordinate and the time, in seconds, on the abscissa. The instrument is stopped after each 5 sec of load and the value of the load is recorded. When the gel breaks, the measurement is complete. Figure 12-3 indicates that the load curve of an incompressible body forms a straight line with an angle of more than 65° towards the abscissa, while the agar curve forms a straight line with an angle of approximately 40°. The gelatin curve is much lower than that of agar, i.e. for the same measuring time the load of the tactile probe is less, the gel is elastic, more yielding. Generally speaking, the load/time curve in "Gelomat" measurements will be lower for the more elastic gels.

The "Gelomat" has been in constant use since 1971 in the author's own laboratory for the quality control of gel substances and microbiological nutrients. By mid 1975 more than 10,000 measurements of agar and gelatin-containing gels had been made. Tables 12-V to 12-VIII list some examples of measuring results of

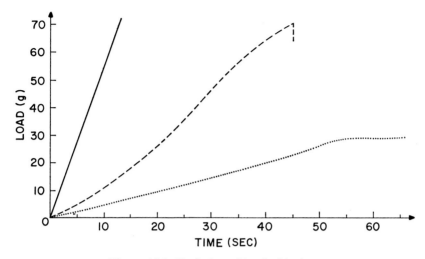

Figure 12-3. Variation of load with time.
———— Incompressible body,
– – – – Agar gel,
. . . . Gelatin gel.

gel stabilities.

Table 12-V illustrates the dependence of gel rigidity on the agar concentration. As expected, the gel stability is directly proportional to the agar concentration of the gel. It should be noted that the measuring time does not rise linearly with the gel rigidity value, i.e. the elasticity of the gel is reduced as the agar concentration increases. Table 12-VI illustrates that the gel rigidity decreases as the pH decreases. It is evident from the table that the greatest rate of decrease occurs between pH 5.5 and 5.0. This fact is important in the manufacture of acid agar-containing nutrients. No noticeable increases in gel stability are to be expected above pH 7.2.

Table 12-VII shows the dependence of gel rigidity of a gelatin-containing gel on the gelatin concentration. Note the relatively long measuring time caused by the elasticity of gelatin gels. Table 12-VIII illustrates some examples of gel stability measurements of various batches of Merck dehydrated media. According to quality control standards, only batches having a gel rigidity between 50 and 80 g are admissible.

TABLE 12-V

DEPENDENCE OF GEL RIGIDITY ON CONCENTRATION AGAR MERCK #71617, STANDARD MEASURING METHOD (VARIOUS EXPERIMENTAL MEDIA).

Measurement No.	Concentration (%)	Gel rigidity (g)	Time (sec)
1-3	1.2	64	35
		66	36
		66	43
4-6	1.5	84	41
		90	45
		90	41
7-9	1.8	114	44
		118	45
		120	45
10-12	2.0	137	47
		134	46
		138	48
13-15	2.5	172	51
		168	49
		172	53

TABLE 12-VI

DEPENDENCE OF GEL RIGIDITY ON pH VALUE
(AGAR, MERCK #1614, STANDARD MEASURING METHOD)

pH	Gel rigidity (g)	Time (sec)
5.0	19	29
5.5	49	45
6.0	52	56
6.5	55	56
7.0	58	46
7.5	59	53

TABLE 12-VII.

DEPENDENCE OF GEL RIGIDITY OF GELATIN ON CONCENTRATION
(GELATIN MERCK #4070, STANDARD MEASURING METHOD)

Concentration	Gel rigidity (g)	Time (sec)
5	21	51
7	24	69
9	60	84
10	68	86
12	82	87

TABLE 12-III

GEL RIGIDITY OF VARIOUS DEHYDRATED MEDIA (MERCK), STAN-
DARD MEASURING METHOD: 3 DIFFERENT TEST BATCHES OF
THE SAME NUTRIENT.

Nutrient	Gel rigidity (g)	Time (sec)
China Blue-Lactose Agar sample	60	35
Sample No. 2348	70	36
	70	38
Endo-C Agar	49	43
Sample No. 4044	69	47
	62	43
MacConkey Agar	58	38
Sample No. 5465	53	39
	54	39
Brolac Agar	50	32
Sample No. 1639	47	28
	43	30
Antibiotic Agar No. 1	53	27
Sample No. 5272	48	29
	48	28
PKU Test Nutrient	65	24
Sample No. 5274	64	27
	62	34

REFERENCES

Barry, A.L. and G.D. Fay: 1973. The amount of agar in antimicrobic disk susceptibility test plates. *Am J Clin Path, 59*:196.

Bauer, A.W., W.M.M. Kirby, J.C. Sherris, and M. Turck: 1966. Antibiotic susceptibility testing by a standardized single disk method. *Am J Clin Path, 36*:493.

Bechtle, R.M. and G.H. Scherr: 1958. A new agar for *in vitro* antimicrobial sensitivity testing. *Antibiot Chemother, 8*:599.

British Standard Specification. No. 757 Methods of testing gelatins. 1944.

Brock, T.D.: 1957. A method for studying antibiotic diffusion in agar. *Antibiot Chemother, 7*:243.

Campbell, L.E.: 1938. The calibration of jelly-testers. *J Soc Chem Ind* (London), *57*:413.

Chakraborty, D.: 1948. On the setting property of agar-agar. *Q J Pharm Pharmacol, 21*:159.

Drew, R.B.: 1946. The determination of jelly strength in absolute units. *J Oil Colour Chem Ass, 29*:259.

Ericsson, H.M. and J.C. Sherris: 1971. Antibiotic sensitivity testing. Report of an International Collaborative Study. *Acta Pathol Microbiol Scand [B.] Suppl, 217*:1.

Frank, H. und F. Buchmayer: 1961. Zur frage des einflusses von agar auf die keimzahlung. *Milchwiss, 6*:302.

Fulthrope, A.J.: 1951. The variability in gel-producing properties of commercial agar and its influence on bacterial growth. *J Hyg, 49*:127.

Goring, D.A.I.: 1956. A semimicro gelometer. *Can J Technol, 34*:53.

Grove, D.C. and W.A. Randall: 1955. *Assay Methods of Antibiotica: A Laboratory Manual.* Medical Encyclopedia, Inc., New York.

Hanus, F.J., J.G. Sands, and E.O. Bennett: 1967. Antibiotic activity in the presence of agar. *Appl Microbiol, 15*:31.

Heicken, K.: 1948. Uber die verwendbarkeit und Eigenschaften säurebehandelter kartoffelstärke zur herstellung fester Nahrboden. *Zentralbl Bakteriol [Orig], 152*:344.

Jones, N.R.: 1956. A tentative method for the determination of the grade strength of agars. *Analyst, 81*:243.

Kizevetter, J.V.: 1937. Fish oceanography, Vladivostock. *Bull Pacific Sci Inst, 37*:5.

Lockwood, H.C. and R.S. Hayes: 1931. A new method for testing agar and gelatin jellies. *J Soc Chem Ind, 50*:145.

Sheppard, S.E., S.S. Sweet, and J.W. Scott: 1920. The jelly strength of gelatins and glues. *J Industr Engineer Chem, 12*:1007.

Yousef, R.T., M.A.E-Nakeeb, and M.H.E-Masri: 1967. Binding of antibiotic to agar and its relations to the inhibition zone. *Acta Pharm Suec, 4*:253.

Chapter 13

THE BIODILUTOR AND BIOREACTOR FOR AUTOMATED TESTS ON SERIAL DILUTIONS OF SAMPLES

E. ENGELBRECHT

Abstract

The practical problems encountered in the execution of analyses requiring serial dilution of samples, and the addition of one or several reagents, are discussed in view of the need for an apparatus allowing individual samples to be examined economically by reliable quantitative methods. Two machines, one an economic table top item, the other a fully automatic robot-technician, are described. Both employ a multichannel, high-precision, microdosage, peristalic dispenser especially developed for this purpose and a silent mixing system. Both machines can distribute a number of samples, dispense a number of fluids simultaneously in predetermined doses, prepare stepwise-increasing dilutions of the samples without splashing, add the diluent in the same operation, dose a number of reagents to any of the rows of receivers, prepare the required control tests, add another series of reagents in a second working-cycle, and mix the reacting fluids. Any program can thus be executed rapidly and automatically with an accuracy and a reproducibility of 99 percent. The diluting system is self-cleaning. The machines can be readied for use and cleaned up in a few minutes. Maintenance can be done by the operator. Because of their versatility they should find many applications, particularly in serology, microbiology, biochemistry, toxicology, and pharmacology.

INTRODUCTION

T HE growing need for automation of laboratory investigations is best demonstrated by the requirements for the control of milk and related products, which entails the epidemiological survey of cattle, the bacteriological control of the milk, and the determination of toxic substances, antibiotics, and disinfectants it may contain. The epidemiological survey of cattle is generally accomplished by way of serological tests, performed either on blood samples taken from the animals or on the milk. The bacteriological control of the milk requires the enumeration of microorganisms present in representative dilutions. Antibiotics and disinfectants are often determined by adding a known number of selected, sensitive microorganisms to increasing dilutions of the milk, followed by incubation and subsequent observation for inhibition of growth. Classical manual techniques are time-consuming, and these analyses are often reduced to single dilution tests performed on pools of samples. It would be preferable to analyse each sample by reliable quantitative methods. This will only be economically feasible with the technical assistance of a machine capable of accurately preparing stepwise increasing dilutions of the samples and submitting them to the required analyses, as well as to the individual control tests, in such a way that cross contamination between samples, accidental contamination, foaming, and aerosol formation are avoided. Such an apparatus should preferably be a multichannel machine, in order to increase capacity and allow comparative tests to be run simultaneously under identical conditions. It should be programmable and sufficiently flexible to allow, for example, tests to be run in duplicate, or analyses of a different nature to be performed successively. Furthermore, it should be possible to equip the machine with a display. It should be composed of easily accessible units to ensure their quick replacement in case of failure. The diluent and reagent dispensing systems should be easily removable and withstand autoclaving.

The technical conception of such an apparatus should be simple enough for the average operator to understand and ensure reliable operation and easy maintenance. It should take little time to prepare the machine for use, to clean it afterwards, and to change from one program to another. Its application should not need any particular skill or technical knowledge. In this paper two patented devices [manufactured by BIOTECHNIA, POB 562, Voorburg, Holland] are described. One of these, named Biodilutor®, is a semiautomatic, economic table top item; the other one, called Bioreactor®, is a fully automatic robot-technician.   Both these machines employ a new, patented type of multichannel, high precision peristaltic dispenser, the Biopulsor®.

TECHNICAL DATA
The Biopulsor

The Biopulsor is the basic element of both the Biodilutor and Bioreactor.   It carries twelve autoclavable silicone-rubber tubes in the case of the Biodilutor, and sixteen in the case of the Biore-actor (Fig. 13-1).

The Biopulsor is a peristaltic pump with a commutator mounted on the rotor shaft; this ensures the accurate delivery of predetermined doses of fluid, unlike conventional pumps which may deliver slightly smaller or larger doses depending on just where the rollers stop.   By using small bore tubing, microdoses can be accurately delivered.   The tubes are fixed to the notches of

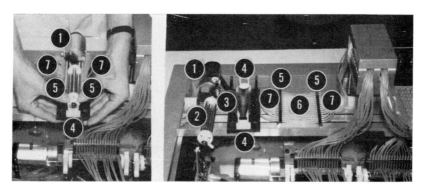

Figure 13-1. The Biopulsor

1. Motor and speed reducer.
2. Three-roller rotor.
3. Pump frame and tube support.
4. Motor, speed reducer, rotor, coupled commutator, and quick-release mount.
5. Quick-release bars.
6. Pump tubes.
7. Connecting tubes.   Tubes and nozzles can be quickly removed from the working post.

two removable bridges and can be rapidly removed for exchange or sterilization.

By reversing it one step, the Biopulsor can be made to draw into each tube a volume of fluid identical to that delivered. One may thus not only distribute, or accurately sample, reproducible volumes of fluid, but also dilute aliquots of liquids in accurately determined dilutions. For example, with the tubes filled with diluent, a predetermined volume of the aliquot is first withdrawn by having the rotor reverse one step. The dilution is then prepared by advancing the rotor two steps, thus delivering the sampled aliquots and adding an equal amount of diluent, flushing the nozzles at the same time. Larger volumes, or higher dilutions, may be obtained by increasing the number of steps proportionally.

The Biodilutor

The Biodilutor has been previously described (Engelbrecht, 1974). Its technical conception and use are briefly explained here. The Biodilutor (Fig. 13-2) is composed of three elements: a twelve-channel Biopulsor, a hand panel, and a case.

The hand panel carries twelve stainless steel nozzles, connected to the channels of the Biopulsor by autoclavable, silicone-rubber tubing, twelve associated mixing devices, a program selection switch, a start button for remote control, and two indicator lamps.

The case holds the Biopulsor. It also contains electric circuitry for manual or automatic operation and an air pump which feeds the mixing devices. It carries a panel on which are mounted the power switch, start/stop switch, and selector switches for manual/automatic operation and for filling or emptying the tubes. The Biopulsor may be fitted with an optional footcontrol for withdrawing or delivering supplementary liquid doses or for multiplying the doses.

Twofold increasing dilutions of the sample may be prepared after completely filling the tubes of the Biopulsor with diluent. If necessary, the tubes may be filled with different diluents. The selector switch is set to "automatic." The nozzles of the hand panel are then introduced into the samples and the Biopulsor is

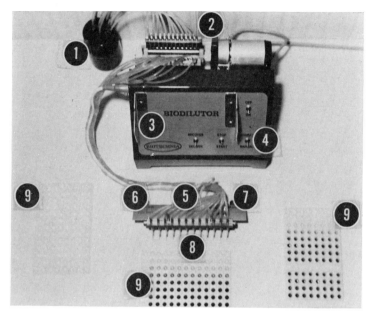

Figure 13-2. The Biodilutor

1. Reagent or diluent vessel.
2. 12-channel Biopulsor.
3. Case.
4. Selector switches.
5. Hand panel.
6. Program selector switch.
7. Start button.
8. Nozzles (12 of) and associated mixing devices.
9. Test trays, containing serial dilutions.

instructed to withdraw a predetermined volume of each sample by depressing the start button. The nozzles are introduced into the first row of receivers, and the cycle of operations is engaged by pushing the start button a second time. The Biopulsor now dispenses the samples and adds an equal volume of diluent. In the meantime, mixing is ensured by the mixing devices, which blow air at an angle to the surface of the fluid, thus inducing a whirling motion. After a pause, during which thorough mixing is achieved, the air pump stops, allowing the liquid to settle, and the

Biopulsor, by reversing one step, withdraws the volume to be transferred to the next row of receivers. The cycle is completed in twelve receivers simultaneously in 9 sec. The nozzles are now moved to the second row of receivers and the cycle of operations repeated by pushing the start button.

After preparing the last dilution, the Biopulsor first ejects fluid removed from the last row of receivers, then flushes the tubes by dispensing diluent through each of the nozzles. If required, the rinse can be repeated. The device is now ready to dilute another series of twelve samples.

Distribution of reagents into the receivers is done by filling the tubes with reagent instead of diluent. If necessary, any of the tubes can be filled with another reagent. The selector switches are set to "manual" and "deliver." Each nozzle dispenses a dose of reagent every time the start button is depressed. Twelve re-agents are distributed simultaneously in three seconds and will be thoroughly mixed with any fluid already persent in the receivers.

Qualitative tests are performed using the foot control. The tubes are filled with reagent. (Again, a different reagent may be conveyed by any of the channels). By pressing the left pedal, the Biopulsor is made to reverse one step, thus establishing a column of air in the nozzles. The nozzles are introduced into the samples and a predetermined volume of each is withdrawn by again pressing the left pedal. The samples are now separated from the reagent(s) by the column of air.

After introducing the nozzles into the test receivers, the right pedal is pressed three times in succession. The Biopulsor thus ejects the sample aliquots, clears the nozzles of air, and distributes doses of the reagent(s), flushing the nozzles at the same time. In the meantime, the delivered samples are thoroughly mixed with the distributed reagents by the mixing devices. If necessary, the nozzles are rinsed, and the apparatus is ready to execute the next series of tests. Twelve qualitative tests can be performed in 18 sec, regardless of whether they are identical or different.

The Bioreactor

The Bioreactor and some of its applications have also been described in earlier papers (Engelbrecht, 1974) and are only briefly described here. The Bioreactor consists of an eight-channel diluting Biopulsor, combined with two sixteen-channel reagent-dispensing Biopulsors, a transport mechanism, a working post, a programming device, automatic commutators, electric circuits for manual or automatic operation, air-pumps, and a variable-speed rotating agitator with spill-tank and flask holders (Fig. 13-3).

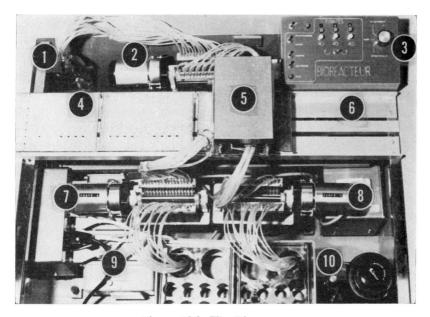

Figure 13-3. The Bioreactor

1. Reagent or diluent vessel.
2. Diluting Biopulsor.
3. Control circuit and case.
4. Test trays.
5. Working post, carrying the outlets of three Biopulsors and two sets of mixing devices.
6. Guide rail for test trays.
7. First-reagent dispensing Biopulsor.
8. Second-reagent dispensing Biopulsor.
9. Air pumps for mixing devices.
10. Variable-speed rotating shaker for reagent vessels.

The test trays, each bearing eight rows of thirteen receivers, are advanced step-by-step along a guide rail. They thus arrive under the working post, which comes down onto the receivers as each test tray stops. The working post is provided with removable bars, one carrying the nozzles of the diluting Biopulsor and the eight associated mixing devices, two others carrying the paired outlets of the reagents dispensing Biopulsors, and a fourth carrying a set of eight supplementary mixing devices. Diluting, dispensing of reagents, and mixing can thus be accomplished in the receivers of the eight rows simultaneously.

In certain tests, two different reagents or pairs of reagents must be added to the various receivers. An example is furnished by the complement fixation test, in which antigen and complement must be added to the analysis vessels, while complement and antigen base solution must be distributed into the control vessel for nonspecific complement inhibition. The availability of two reagent dispensing Biopulsors, each provided with two channels per row of receivers and programmed individually, allows these two operations to be mechanized. In addition, the diluting Biopulsor can prepare the blank control by having it add a supplementary quantity of buffer solution to the corresponding receivers. The reagent storage bottles may be agitated in the variable speed rotating agitator. Reagents can also be cooled during the time of operation by placing ice in the spill tank.

The program consists of a paper sheet on which the pumps are indicated by symbols, and the succession of receivers is indicated in the squares following these symbols. The apparatus is programmed by punching holes in the corresponding squares. The program sheet is wrapped around a metallic cylinder, which is mounted on the spindle moving the test trays and is situated under the box containing the low-voltage control circuit. The bottom of the box is provided with a set of contacts which rest on the paper. Each time the test trays are advanced, a fresh row of program holes is interposed between the contacts and the cylinder. Electrical contact is made wherever a hole has been punched, and the instruction coded by the hole is transmitted to the control circuit. The program sheet carries two programs, a main one controlling all the operational parts of the machine, and an auxiliary one controlling the reagent-distributing Biopulsors.

The three Biopulsors are programmed individually. Two rows of squares on the main program are reserved for the diluting Biopulsor, one for sampling, the other for delivery of the aliquots taken earlier and the addition of an identical quantity of diluent. The diluting Biopulsor may thus pick up a sample from any receiver in the row, transfer it to any further receiver, start preparing a series of dilutions, or stop diluting at any place in the tray, and extend the series over any number of receivers. It can eject fluid removed from the last dilution, flush the nozzle, and rinse its outside, thus cleaning the channels. Therefore, the machine may handle successive samples without stopping. Eight quantitative tests can thus be executed simultaneously according to any program. As each of the channels, or pairs of channels, of the reagents-dispensing Biopulsors can convey a different reagent, these tests can even be of a different nature.

When no dilutions have to be prepared from the aliquots, for instance when qualitative tests are performed, or when ampoules have to be filled, it may be convenient to have a supplementary dispenser. For this purpose the panel of the machine is provided with a selector switch allowing the diluting Biopulsor to be used as a simple reagent dispenser. By punching holes in either one or both of the rows of squares, either one or two doses can then be added to any of the receivers. Each Biopulsor can then dispense a different reagent, or pair of reagents, through the various channels. The Bioreactor can thus distribute twenty-four reagents, or pairs of reagents, simultaneously, in identical or different combinations.

One row of squares is reserved on both the main and auxiliary programs for each of the reagent dispensing Biopulsors. By simply switching from "main program" to "auxiliary program," analyses which must be run in two stages can be completed without waste of time.

RESULTS AND DISCUSSION
Accuracy and Reproducibility

The accuracy and the reproducibility of the metering of doses by the Biopulsor were verified for each of the channels separately by repeatedly dispensing doses into a receiver placed on a precision balance. The doses delivered appeared to be repeatable within 0.5 percent. When the tubes were well cut, at the proper length, an accuracy of 98 percent could be obtained without difficulty.

The accuracy and the repeatability of diluting were verified for each channel by having the machine repeatedly prepare a 1:5 dilution of a stock solution of methyl violet, followed by eight twofold dilutions. For the purpose of verification, the final dilution of 1:1280 was also obtained by mixing 0.5 ml of the concentrated dye solution with 639.5 ml of the diluent. The dilutions were then compared photometrically. The diluting error appeared to be within the range of error of the manual pipetting and the photometrical reading and could thus not be properly defined. However, the cumulative error was found to be less than 1 percent per step, while the error of repeatability did not exceed 0.5 percent.

These results, which are satisfactory for any practical or scientifical application, are confirmed in daily use, as demonstrated by the following example: Forty sera were examined twice for antibodies against the haemolysins of streptococci in the AST, a quantitative, two-step lysis inhibition test. The results obtained in the two series were perfectly identical (Fig. 13-4). The reproducibility of the results was demonstrated by the titration of the autostreptolysin antibodies in twenty-four sera on two Bioreactor machines simultaneously, using reagents from the same stock. Again the results were identical.

The performance of the Biodilutor depends on the technique of the particular test system. For example, 100 qualitative tests can be executed in 3 min. For quantitative tests, depending on the number of twofold increasing dilutions to be prepared, 100 tests can be executed in 20 to 35 min, including the distribution of one reagent into any of the receivers. It takes another 10 min to empty the tubes of the Biopulsor, refill them, and add a second reagent to the 800 to 1200 receivers. In each of these cases, of course, the samples may be submitted to up to twelve different analyses simultaneously, together with the individual control tests.

The performance of the Bioreactor is determined by the duration of the operational cycle and the number of receivers covered

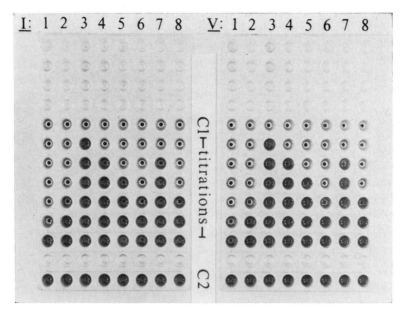

Figure 13-4. Reproducibility test

I   Tray No. 1.   First tray of a series of quantitative determinations of streptococcal haemolysin-neutralizing antibodies on individual serum samples of human origin;

V   Tray No. 5.   First tray of the repeated quantitative antistreptolysin determinations.

1-8   First 8 of 40 sera, examined twice by titration for the presence of streptolysin-neutralizing antibodies by preparing twofold increasing dilutions of the samples, adding buffer solution to the control receivers [absence of disturbing haemolytic factors (Cl)], dispensing identical volumes of streptococcal haemolysin into the dilutions, preparing controls for the haemolytic power of the streptococcal haemolysin (C2), followed, after subsequent incubation, by the distribution into the receivers of a predetermined quantity of a 4% suspension of human 0-group erythrocytes. Note the similarity of the degree of haemolysis obtained in the corresponding rows of the two trays.

by the program.   The operational cycle takes 16 sec, leaving 4 min for the operator to distribute aliquots into the next tray and to do the accompanying clerical work.   An efficient interaction is thus ensured between the operator and the machine.   The machine handles 15 test trays in 1 h.   Thus, 240 semiquantitative tests or 120 full titrations can be performed, with the simultaneous addition of one to four reagents as required, and completed with the individual control tests, every hour.

While the Bioreactor works somewhat slower than the Biodilutor, the technician may use the extra time to distribute aliquots and deal with paperwork.   Because the Bioreactor executes the program automatically, human errors are excluded, even when nonprofessional staff carry out the analyses.

Manipulation and maintenance, being very simple and self-explanatory, require no special professional instruction or particular technical knowledge.   As the machines can be readied for work and cleaned afterwards in a few minutes, they can also be employed to advantage in small series of tests. Simplicity of mechanical conception ensures a reliability and allows maintenance to be done by the operator himself.

REFERENCE

Engelbrecht, E.: 1974. The Bioreactor and Biodilutor as tools for the automation of tests to be performed on serial dilutions of samples.   IAMS—IX. *Int Symp Kiel,* Abstracts 35.

Chapter 14

PROGRESS IN IMPEDANCE MEASUREMENTS IN MICROBIOLOGY

P. CADY

Abstract

A novel approach to performing the microbiological tasks of detection, enumeration, identification, and determination of antimicrobial susceptibility is presented. The relationships between impedance changes and microbial growth and metabolism is reviewed. Several commercially available impedance measuring instruments are described. Progress in the application of impedance measurements for enumerating bacteria in clinical urine specimens and food samples is described. The detection of molds and yeasts is discussed, as is the detection and identification of mycoplasma. The potential of impedance measurements for microbial antibiotic susceptibilities, microbial characterization, and identification is also presented. The impedance approach is rapid, inexpensive, frequently requires little or no sample preparation, works well with opaque samples, is convenient to use, and is commercially available.

INTRODUCTION

CURRENT interest in the use of impedance measurements as a tool of general use to the microbiologist for detection, enumeration, and characterization of microorganisms stems from papers presented at the First International Symposium on Rapid Methods and Automation, held in Stockholm in June of 1973 (Cady 1975, Ur and Brown, 1975b). More recently, Hadley and Senyk (1975) reviewed their initial clinical experimentation with automated impedance measuring instruments. This

---

*The author thanks S. Dufour, S. Kraeger, R. Mischak, K. Hadley, D. Hardy, P. Lawless, J. Shaw, E. Dana, and P. Berkenkotter for their experimental work, and S. Dufour, S. Kraeger, A. Cavette, J. Wong, B. Brindley, and L. Yutzy for their assistance in preparing the manuscript and illustrations.

199

paper will review some of the more recent progress made with impedance measurements not covered in Hadley's review.

Background

Although impedance measurements have been widely used to describe physiological events (Geddes and Baker, 1968), the physical chemical properties of bacterial cell walls (Carstensen et al., 1965; Carstensen, 1967; Carstensen and Marquis, 1968), and viral particles and macromolecules of biological interest (VanderTouw et al., 1973), there are but scattered references to applications for measuring metabolic activity or growth of microorganisms.

G. N. Stewart, an American, presented a paper to the British Medical Association at Edinburgh, in 1898, describing conductivity measurements made on defibrinated blood allowed to putrey, and reported a tenfold increase in conductivity occurring over a twenty-five day period. The apparatus was that of Kohlraush and was of limited sensitivity, thus only daily measurements were reported.

In the late 1920s, Parsons and Sturges (1926 a,b) and Parsons et al. (1929), working for a meat packing firm and using a more sensitive apparatus, investigated the usefulness of conductivity measurements for determining the metabolic activity of clostridia species incubated anaerobically in various broths. They showed good correlation of conductivity increase with ammonia production. Proteolysis by bacteria was measured conductimetrically by Allison et al. (1938) a decade later. Since alternating current was used in all of these early experiments, impedance was actually measured and conductivity was calculated. Furthermore, the sensitivity of the measurements was sufficiently low that changes in impedance could only be meaningfully recorded at intervals of several hours. In 1958, McPhillips and Snow, using a more modern approach, employed an unusual toroidal conductivity cell to follow acid production in milk samples brought about by *Streptococcus lactis*. The cell had the valuable feature of having no electrodes in contact with the fluid being measured, conductivity being calculated from changes in the electrical properties of currents flowing through coils wrapped about opposite sides of the torus. This avoided the nettlesome problem of electrode polarization impedance discussed by Schwan (1968).

Enzymatic activity can be followed conductimetrically and has been reviewed by Lawrence (1971), Lawrence and Morres (1972), and Hanss and Rey (1971 a, b). A conductimetric titrator and analyzer is described by Andreev (1974) for use in the clinical laboratory but has not, to the author's knowledge, been employed to measure bacterial growth and metabolism.

The Coulter Counter, widely used in hematology, is essentially a particle counter and is based upon changes in the conductivity measured across an orifice as particles, differing in conductivity from the conductivity of the suspending fluid, flow through the orifice. Particles can be sized as well as counted by applying pulse height analysis to the voltage being recorded. Growth rates of microorganisms and changes in size during the growth cycle can be readily determined (Harvey et al., 1967). A practical consideration is that to conduct these experiments, which require a very fine orifice, particle-free media must be used to avoid frequent plugging of the orifice and to avoid artifacts in enumeration. Usually, no distinction between viable and nonviable cells can be made.

Mention should also be made of the work of Goldschmidt and Wheeler (1973) and of Wheeler and Goldschmidt (1975) who used a four-electrode impedance-measuring system to quantitatively detect bacteria in urine samples. This approach does not rely upon microbial metabolism and is a static rather than dynamic measure. The bacteria are removed by filtration, washed with distilled water, and resuspended in sterile pyrogen-free water. The impedance of solutions thus prepared is observed to increase with increasing numbers of bacteria in the range of 10^3 to 10^9 organisms per ml. A square wave pulse train from 1 to 100 Hz frequency and up to 20 V (peak-to-peak) was used. Best results were obtained at 10 Hz and 10 V (peak-to-peak). The method is rapid and correlates with simultaneous plate counts to within 5 percent. The size and shape and surface characteristics of bacteria appear to be without effect on the relationship of bacterial concentration to impedance, nor does viability play a role (personal communication).

Cady (1975) and Ur and Brown (1974, 1975a,b), working independently, described the use of continuous impedance measurements to follow microbial metabolism over time periods of several minutes to several hours. Both investigators employed two identical impedance measuring chambers, both filled with identical medium, only one of which was inoculated, the other serving as a reference or control. This markedly reduced the contribution to impedance changes brought about by fluctuations of temperature, chemical changes in the medium, electrical noise, absorption of gases, etc.

Impedance

Impedance is the resistance to the flow of a sinusoidal alternating current through a conducting material. It is a complex entity, consisting of a resistive component R and a reactive component X. For our discussion, we will ignore any contribution to the reactive component from inductance, which plays no discernible role in our measurements and assume that the reactance is only due to capacitance, designated as X_c or capacitive reactance. If one plots time against the alterations in voltage and current when a circuit consists only of an alternating current passing through a resistor, it is seen that the voltage and current rise together to a peak, first in one direction and then in another. The current and voltage are said to be in phase. In a capacitor, however, the maximum flow of current occurs when the voltage is at zero and vice versa. The current flow is said to lead the voltage by a quarter of a cycle or to be out of phase by 90°. The relationship between voltage E and current I for a given frequency f, when a circuit contains both a resistor R and a capacitance C in series is given by:

$$E = I \sqrt{(R^2 + X_c^2)}$$

where X_c is the capacitive reactance and is given by:

$$X_c = \frac{1}{2 \pi f C}$$

The term $\sqrt{(R^2 + X_c^2)}$ is the impedance and is designated by the term Z.

The phase angle Φ is given by:

$$sin\ \Phi = \frac{X}{Z}\ \ or\ \ cos\ \Phi = \frac{R}{Z}.$$

These relationships can be readily seen in Figure 14-1. For further discussion see Stacy (1960).

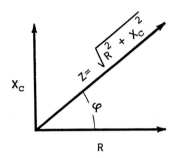

Figure 14-1. Representation of impedance as the vector sum of the resistance and capacitive reactance.

Impedance Changes During Microbial Growth

The microbiological test system consists of a pair of electrodes placed into a growth medium into which microorganisms or other cells have been inoculated. An alternating current is then passed from one electrode through the medium to the other electrode, and the impedance to this flow is measured. Electrolytes in solution conduct current by the transport of ions and charged molecules, in contrast to electronic conductors in which electrons are the charge conductors. Thus, through the bulk of the solution, factors which affect the mobility of the ions such as temperature, concentration density, molecular weight, and transference number in turn affect the measured impedance. At the surface of the electrodes, layers of charged molecules, attracted to or bound to the electrode's surface, represent stored charges and cause the surface to act as a capacitor. We can represent the electrode-medium interface by a model which consists of a resistor and a

capacitor in series (Fig. 14-2). This describes, as a first approximation, the electrical circuit represented by the test system, and
is essentially that described by Geddes et al. (1971) and Schwan
(1963). See also Warburg (1899, 1901). If we assume that the
capacitance of the medium is small relative to that at the electrode, then the system can be represented by a circuit with two
capacitors and a resistor in series, and the impedance is given by:

$$Z = \sqrt{(R^2 + X_c^2)}$$

where R equals the sum of the resistances at the electrode and of
the medium, and X_c equals the sum of the capacitive reactances
at the electrodes.

During microbial growth and metabolism the resistance is
generally seen to decrease. This is thought to reflect the creation
of ion pairs by metabolic activity such as the conversion of glucose
to lactic acid, the metabolism of fats to bicarbonate, or the in-

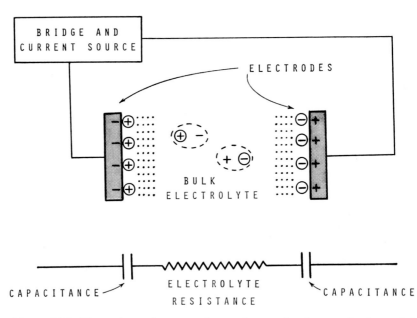

Figure 14-2. The series resistance and capacitance circuit as a simple model
for the effects of electrolyte resistance and double layer capacitance at the
surface of the electrodes.

crease in mobility caused by the cleavage of large charged molecules, such as proteins into smaller, more mobile molecules, such as amino acids. Decreases in resistance typically range from 1 to 12 percent during an 8 h growth experiment.

Also during microbial growth and metabolism, the capacitance is generally seen to increase. This increase is not well understood but may be caused by the adsorption of molecular species possessing a greater dielectric constant than those initially adhering to the electrodes. The capacitances measured are very large, in the range of several microfarads. Generally, the percentage increase in capacitance is equal to or greater than the percentage decrease in the resistance.

Figure 14-3 shows the impedance changes resolved into resistive and reactive components associated with *Escherichia coli* and *Serratia marcescens* both growing in trypticase soy broth (TSB). This was accomplished by measuring each impedance value at two different frequencies, assuming the model of a series capacitor and resistor, and solving the multiple simultaneous equations generated thereby for C and R. This is only an approximation since the electrodes deviate somewhat from this model. Similar deviations are described for stainless steel electrodes by Geddes et al. (1971).

Note that for both organisms the percentage decrease in the capacitive component X_c is much larger than in the resistance component R. For most of the organisms studied, the contribution to the impedance change from the capacitive reactance is much greater than the contribution from resistance. Since the capacitive reactance is a function of the reciprocal of the frequency, the impedance change is seen generally to be much larger at lower frequencies. This is illustrated by Figure 14-4 where the impedance change produced by *E. coli* growing in brain heart infusion broth (BHI) is shown at different frequencies.

The magnitude of the impedance change observed is a function of the microorganism under study, the medium in which it is growing, the frequency of the signal applied, the surface properties of the electrode used, the geometry of the electrodes, surface to volume ratios, and interelectrode distances. The

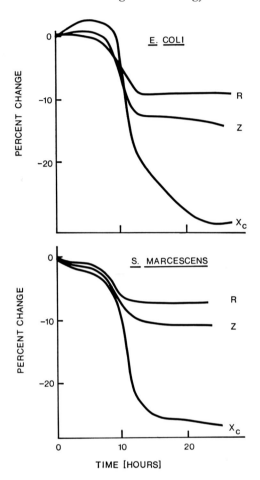

Figure 14-3. Changes in resistance (R), capacitative reactance (X_c), and impedance (Z) observed with *Escherichia coli* and *Serratia marcescens*. Values were deduced from measurements at 400 Hz and at 2 KHz using the model shown in the previous figures. The organisms were grown in TSB with 0.1% agar and monitored with stainless steel wire electrodes.

greater part of the impedance change measured is due to surface effects on the electrodes and, to a lesser degree, changes in conductivity of the bulk medium between the electrodes.

In general, undefined media rich in protein such as brain heart

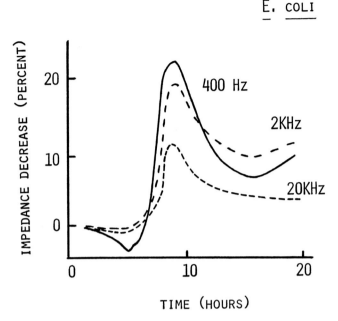

Figure 14-4. Impedance changes resulting from *E. coli* (10³ per ml inoculated in BHIB) observed at three different frequencies.

infusion broth or trypticase soy broth give greater changes in impedance for a given amount of metabolism than chemically defined media. Electrode surfaces which have given good responses to growing microorganisms include gold, stainless steel, chrome, silver, and tin-nickel. Nickel, nickel-boron, and copper electrodes have not been satisfactory. It should be noted that although excellent responses are achieved with gold plated electrodes, there is a great deal of variation between electrodes prepared under gold plating conditions varying only slightly with respect to gold concentration, current density, and surface preparation. Similarly, stainless steel electrodes vary widely in sensitivity to microorganisms and uniformity of response, depending upon small variations in the conditions of surface preparation.

Impedance Measuring Instruments

Bactometer® 32 Microbial Monitoring System

To apply impedance measurements to microbiological tasks, we have used an exceptionally sensitive impedance bridge called the Bactometer 32, manufactured by Bactomatic, Incorporated of Palo Alto, California (Fig. 14-5). To make measurements, a cluster of impedance chambers arranged in pairs called a "module" or series of bottles filled with electrodes arranged in pairs are employed. Typically a module consists of sixteen

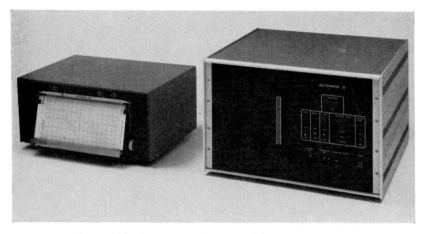

Figure 14-5. Bactometer 32 and multichannel recorder.

chambers, each of 2 ml volume, and each fitted with impedance electrodes. Both chambers (or bottles) of a pair are filled with identical medium. One chamber of the pair is inoculated with the sample to be tested, the uninoculated chamber serves as a reference. The module is then inserted into the incubator portion of the instrument, where electrical connection is made between the impedance measuring circuitry of the instrument and the chamber electrodes. The instrument now automatically measures the ratio of impedance in the reference to impedance of both reference and sample as given by the following relationship:

$$r = \frac{Z_{ref}}{Z_{ref} + Z_{sample}}$$

Expressing the data in terms of impedance ratios has the advantage that alterations in impedance occurring in the medium of both chambers, such as changes due to temperature, absorption of gases, and chemical changes, have no effect upon the ratio. Furthermore, the ratio when both chambers are identical is one-half. Increases in impedance of the sample cause the ratio to decrease, and vice versa. Prior to inoculation the ratio should be very close to one-half. Generally, inoculation of a 1 ml sample with 10 to 100 μl has little effect on the ratio, but even if it does, the succeeding changes in ratio are reasonably independent of the initial values.

The Bactometer 32 passes a small current of 40 mV potential across both the chambers, reference and sample (this is approximately 20 mV across each chamber) at a frequency of either 400 Hz or 2,000 Hz. The instrument then measures the impedance ratio of each of thirty-two pairs of chambers. Measurements are taken every 3 sec; thus all thirty-two samples are sequentially measured every 96 sec.

The impedance ratio is digitized into a sixteen-bit binary number, using an A/D convertor of the author's design. This binary number is converted into a decimal number for display on the front panel, converted to an analog voltage for display on a strip chart recorder, or fed directly together with a sample address to a computer by way of a serial interface. The Bactometer 32 is able to selectively monitor a single chamber pair or a single module (cluster of eight chamber pairs). When used in the single chamber mode, data points are accumulated every three seconds, enabling one to record more rapidly changing data, such as from enzyme kinetics.

The Bactometer® 8, Microbial Monitoring System

This is a smaller and simpler instrument handling up to eight samples at a time (Fig. 14-6). This instrument presents its data on the front panel or on a strip chart recorder as does the Bactometer 32, but in addition has a printer option which allows data to be printed out every 15 min or every 2 h. It does not have its own incubator and the impedance measuring circuitry employs a ten-bit A/D convertor in contrast to sixteen bits for the Bactometer 32. This reduced sensitivity is of importance for some organisms and some media which do not produce enough impedance changes to be detectable.

The above instruments produce digital data outputs in binary form. This suggests that a larger instrument could readily be made, employing a built-in digital computer to handle micro-biological procedures in a completely automated fashion, in accordance with preprogrammed instructions.

The Bactobridge®

This instrument was originally designed for use in blood coagulation studies (Ur, 1970), but more recently has been put to microbiological use by Ur and Brown (1974, 1975a,b). The instrument was originally available from Stratton and Company (Medical) Ltd., (as the Strattometer), but can now be obtained from T. E. M. Sales Ltd., Crawley, Sussex, England, or from Koniak and Partners Ltd., Geneva, Switzerland. Three versions have been available — a single channel, a three channel, and a six channel instrument. All instruments employ pairs of glass capillary tubes with electrodes gold plated in annular fashion at each end, and contain about 100 μl of either sample or reference medium. Great care is taken to assure that each set of electrodes is carefully matched with respect to resistivity, capacitance, and thermal properties. A 10 KHz signal, of less than 0.5 V, is applied

Figure 14-6. Bactometer 8 and multichannel recorder.

across both cells. The bridge is balanced by turning a ten-turn 1K potentiometer. The voltage is recorded on a strip chart recorder as the bridge becomes unbalanced by changes in the impedance of the sample cell. As few as 3×10^5 organisms have been reported to be detected promptly with this sensitive microsystem.

The Titr-2

This is a semiautomatic conductometric titrator, described by V. S. Andreev (1974). To this author's knowledge, this instrument has not been used in microbiological applications.

The Torrymeter

This instrument was devised at the Torry Research Station to measure the freshness of fish, which has been shown to be related to the dielectric properties of fish muscle and skin. The instrument measures a factor which is defined by:

$$\frac{1}{2\pi f C R}$$

where f is the frequency of the measuring signal (2 KHz) and C and R are the capacitance and resistance of the tissue. Fish freshness is influenced by a number of factors including microbial degradation and autolysis of tissue cells and is generally measured by trained fish inspectors who rely upon their sense of smell and taste. This author is unaware of any more direct measurement of microbial activity by this instrument. The instrument uses two pair of concentric electrodes, the outer electrode made of graphite, the inner of stainless steel. Less than 1 mA current is used. The instrument is commercially available from G. R. International Electronics, Ltd., Almondbank, Perthshire, Scotland.

Electrical Safety

Geddes et al. (1969) has referred to the electrical hazards of impedance measurements made at low frequencies when transthoracic electrodes are used to monitor cardiac output or respiration. He was able to estimate current densities for various frequencies necessary to produce sensation, vagal slowing, and even fibrillation, and showed that at higher frequencies larger currents could be tolerated than at lower frequencies.

Electrode potentials used in impedance measurements for microbiological purposes are generally of such low voltage that even if large transthoracic electrodes (with electrode jelly) were inadvertently connected to the instrument and placed in contact with the operator, the current delivered would be far below the level of perception and many fold below physiologically dangerous levels. Thus, the use (and even the misuse) poses little electrical hazard to the operator.

Impedance Chambers

Bottles and Tubes

A variety of impedance chambers have been devised. The simplest of these consists of a pair of bottles or tubes with overhead stainless steel wire electrodes which dip into the solution contained therein (Fig. 14-7). Any volume bottle or tube can be used. The author has had experience mainly with bottles of from 50 to 200 ml capacity and tubes of from 4 to 30 ml capacity. The bottle stopper electrode assembly is fabricated of natural rubber of the sort used for serum stoppers. Thus, these bottles find use in blood cultures and anaerobic cultures. Generally, the electrodes are fabricated of stainless steel and are disposable. This insures a fresh sensitive electrode surface for each experiment or analysis.

Etched Electrode Modules

A module consisting of a cluster of sixteen impedance chambers, arranged in pairs, can be fabricated using a circuit board and a plastic superstructure. The circuit board forms the floor of all the chambers and has etched upon it interdigitating gold plated or stainless steel electrodes together with the necessary leads to conduct the signal from the connector edge of the circuit board to each chamber. Bonded to the circuit board is a plastic superstructure formed into sixteen chambers, each chamber being

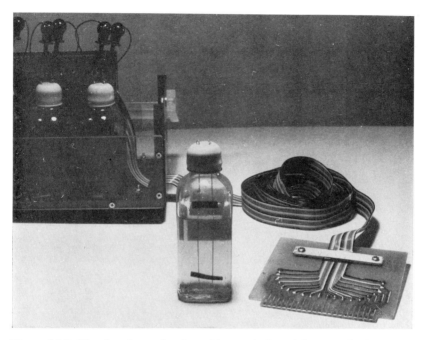

Figure 14-7. Blood culture bottle with vertical stainless steel electrodes. Bottles are placed in a rack incubated outside the Bactometer and connected to the Bactometer with an extension cable.

directly over each of the sixteen pairs of electrodes on the board. Individual caps serve to preserve sterility and prevent evaporation of the solutions which are pipetted directly into each chamber and are in direct contact with the etched electrodes on the board (see Fig. 14-8). To connect this type of module into the instrument, the edge of the module is simply inserted into an edge card connector within the incubator portion of the instrument.

Disposable Modules

Since electrode surface changes occur with each use, it was deemed advisable to make a disposable module so that a fresh electrode surface could be employed for each measurement. These are made by incorporating a stainless steel lead frame into an injection molded part, such that each chamber has a pair of

Figure 14.8. Printed circuit board module with horizontal electrodes.

stainless steel prongs rising up from the floor of the chamber and projecting into any solution contained therein. The remaining portion of the stainless steel is imbedded in the plastic and serves to conduct the current to the edge of the module where it can fit into an edge connector within the Bactometer. The plastic used is styrene acrilonitrile copoloymer, or polycarbonate. Sterilization is accomplished by ethylene oxide or radiation. These disposable modules are available from Bactomatic, individually wrapped and sterilized (see Fig. 14-9).

Bottle Baskets

To connect the bottle electrodes to the instrument, a bottle basket is convenient (Fig. 14-7). The bottles are individually connected to the basket, and the basket in turn is placed in an incubator. Electrical connection to the Bactometer is provided by a flat cable fitted with a connector card at its end, which inserts into the Bactometer. A flat cable allows the incubator door to close without heat loss. The flat cable approach also allows modules of either the circuit board design or disposable design to

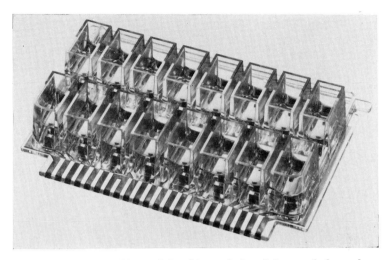

Figure 14-9. Disposable module with vertical stainless steel electrodes.

be placed, at a distance from the Bactometer, in CO_2 incubators, anaerobic chambers, refrigerators, shakers, etc.

RESULTS

Having described the Bactometer, the method of using it and the subsequent results will now be discussed. A module is selected which has been cleaned and sterilized. Media are placed in all the chambers, care taken to insure that the same medium is in the reference chamber for each corresponding experimental chamber. The module is capped and allowed to equilibrate, preferably for several hours with the module plugged into the instrument so that one is sure that there is no impedance change with time, or only slight drift. Experimental wells are then inoculated with the sample and the module reinserted into the instrument.

Evidence that impedance changes do reflect metabolic activity and bacterial growth is shown in Figure 14-10, which depicts the growth of *Salmonella enteritidis* growing in BHI. The number of organisms was determined by sequential sampling and replicate plate counts. The parallelism between microbial growth curves and impedance changes cannot always be assumed, as this rela-

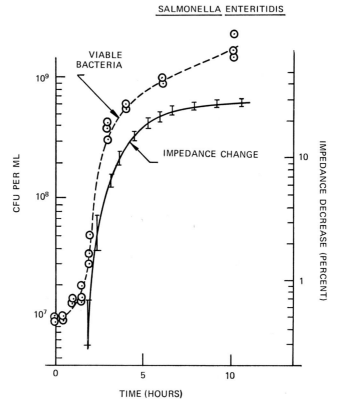

SALMONELLA ENTERITIDIS

Figure 14-10. Comparison of impedance change and bacterial growth curve for *Salmonella enteritidis* growing in BHIB. The circular data points represent plate count results. The bars on the impedance curve represent standard deviations over six replicate channels.

tionship is markedly dependent upon both medium and electrode surface properties.

When the inoculum contains a small number of microorganisms no change in impedance is detected until the organisms have replicated to a threshold concentration. Thresholds for most bacteria, yeasts, and molds growing in most media are about 10^6 to 10^7 organisms per ml. A reduction in the threshold concentration can sometimes be achieved by brief fiber dialysis of the medium to reduce its ionic content, which makes the small contribution of bacterial metabolites to conductivity more readily apparent. The author has observed, as have Ur and Brown (1975a), that a reduction in the buffering capacity of media increases the impedance change for a given number of organisms.

Detection time is clearly a function of the initial concentration of organisms, their generation time, and the threshold characteristic of the medium-organism-electrode combination. Knowing the threshold and the generation time, one can readily calculate the number of organisms initially present. Figure 14-11 depicts the impedance changes observed when three concentrations of *E. coli* are grown in E. C. medium. Note that the highest concentration produces impedance changes most rapidly, followed by the successively lower concentrations of organisms. The time between curves produced by different concentrations of organisms represents the time required for the organisms to replicate from lower to higher concentration, in this case to replicate a hundredfold. From this, one can readily estimate the generation times. Figure 14-12 depicts typical detection times plotted against cell concentration at inoculation. The cell concentration for im-

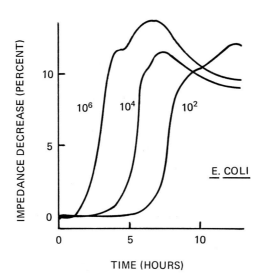

Figure 14-11. Impedance changes resulting from *E. coli* inoculated in ECB in concentrations of 10^6, 10^4, and 10^2 per ml.

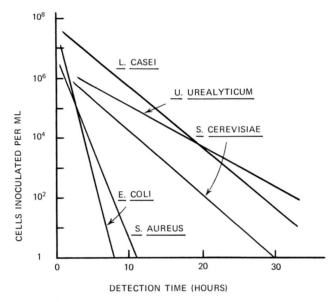

Figure 14-12. Detection times of impedance changes against initial concentration of microorganisms for *E. coli* in TSB, *S. aureus* in TSB, *Saccharomyces cerevisiae* in GPYE Broth, *U. urealyticum* in urea medium (Lawrence, 1971), and *L. casei* in universal beer broth.

mediate detection is defined as the threshold concentration and is seen to lie between 10^6 and 10^8 organisms per ml.

Table 14-I shows threshold values and generation times for a number of different microorganisms in varying media.

DISCUSSION

We have described a simple system for the detection and enumeration of microorganisms. There are several practical applications of such a scheme with which we have had experience.

Single Chamber Measurements

Urine Culture

Estimating the degree of bacteriurea in clinical urine samples has been accomplished at Bactomatic. Results from the first 609 specimens are shown in Figure 14-13. This work was done using clinical specimens from Kaiser Foundation Hospital-Santa Clara,

TABLE 14-I

TYPICAL GENERATION TIMES AND THRESHOLDS FOR A VARIETY OF MICROORGANISMS*

Organism	Medium	Generation Time	Approximate Threshold
Bacteria			
Escherichia coli	TSB	20 min	2×10^7/ml
Proteus vulgaris	TSB	20 min	4×10^6/ml
Salmonella enteriditis	BHIB	20 min	10^7/ml
Pseudomonas aeruginosa	BHIB	25 min	5×10^6/ml
Staphylococcus epidermidis	BHIB	20 min	9×10^6/ml
Staphylococcus aureus	BHIB	35 min	4×10^6/ml
Streptococcus pyogenes	Todd Hewitt Broth	60 min	4×10^6/ml
Lactobacillus brevis (32°)	Universal Beer Broth	90 min	5×10^7/ml
Pedicoccus cerevisiae (32°)	Universal Beer Broth	240 min	5×10^7/ml
Neisseria gonorrhoeae	Thayer Martin Broth	45 min	2×10^7/ml
Fungi			
Saccharomyces cerevisiae (32°)	GPYE† Broth	90 min	10^6/ml
Candida albicans (32°)	TSB + 2% glucose	45 min	4×10^6/ml
Aspergillus niger (32°)	GYPE Broth	70 min	10^7/ml
Mycoplasma			
Acholeplasma laidlawii	PPLO Broth	210 min	10^7/ml
Ureaplasma urealyticum	Urea Medium	150 min	2×10^6/ml

*Unless otherwise indicated, all organisms were grown without agitation at 35°C. All organisms were monitored with gold-plated printed circuit board electrodes except *C. albicans*, which was monitored with stainless steel electrodes.

†Glucose-Peptone-Yeast Extract.

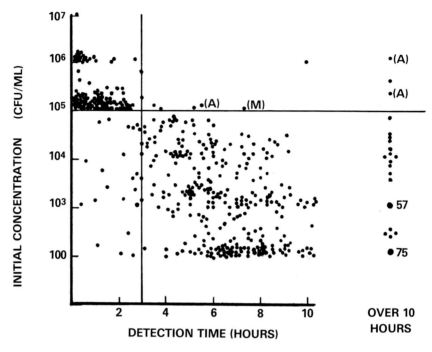

Figure 14-13. Detection times of impedance changes against initial concentration of microorganisms, for 609 clinical urine specimens. The horizontal line at 10^5 organisms per ml indicates the level usually chosen to divide positive and negative specimens. The vertical line at 3 h indicates the optimal cutoff time for the impedance-based test. Numbers adjacent to larger black circles indicate the number of samples having the same detection time and initial concentration, displayed together as one large data point. (A) indicates samples from patients who were already being treated with antibiotics. (M) indicates a mixed culture.

and was presented by Dr. Spring Kraeger at the Annual Meeting of the American Society for Microbiology in 1975. The graph shows a scatter plot of detection time (in hours) versus the number of organisms as determined by plate count. Of particular interest to the clinician is the ability to distinguish urine samples containing greater than 10^5 organisms per ml, since Kass (1955) has shown that, in a population of patients, 96 percent of those with urinary tract infections had urine specimens containing 10^5

or greater microorganisms per ml, whereas only 4 percent had urine specimens containing less than 10^5 organisms per ml. Thus, using Kass' guideline, we define a positive culture as one containing in excess of 10^5 organisms per ml and a negative culture as one containing less than 10^5 organisms per ml. To separate positive and negative as thus defined, a 3 h detection time was used. Cultures detected in under 3 h were positive, and over 3 h were negative, by definition. The correlation between cultures defined as positive by Bactometer detection and by plate count method was 94 percent, well within the error usually associated with the plate count method. This series has been recently extended to over 1,100 cultures with similar results. Table 14-II indicates the organisms isolated from the positive cultures in this initial series of 609 cultures. Of particular interest is the observation that over one-half of the positive cultures were detected within 1 h.

TABLE 14-II

DISTRIBUTIONS OF ORGANISMS IN POSITIVE CULTURES

Organism	Number of Terms Isolated	Percent
Escherichia coli	60	49.2
Proteus mirabilis	10	8.2
Klebsiella pneumoniae	7	5.7
Staphylococcus aureus	5	4.1
Enterococcus	5	4.1
Pseudomonas aeruginosa	4	3.3
Serratia marcescens	1	0.8
Enterobacter agglomerans	1	0.8
Streptococcus sp	1	0.8
Candida sp	1	0.8
Unidentified Gram-negative rods	18	14.8
Unidentified Gram-positive cocci	4	3.3
Mixed	2	1.6
Total	122	

Figure 14-14 depicts the relationship between false negative and false positive cultures, as determined by the plate count method, and the cutoff time selected to represent 10^5 organisms

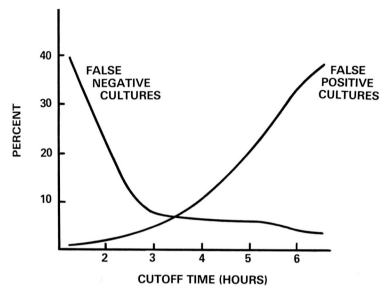

Figure 14-14. Percentages of 609 urine cultures which would have been falsely classified if various cutoff times were used.

per ml. Note that the false negative cultures do not drop very rapidly after 3 h, whereas the false positive cultures rise quite rapidly after 3 h. Therefore, cutoff times of 3 h or less are recommended for this kind of screening procedure. A screening test such as this gives rapid information to the clinician and may prove helpful in establishing therapy protocols based on more than clinical symptoms.

Frozen Vegetables

Another similar application is in establishing acceptable or unacceptable levels of bacterial contamination in frozen vegetables. Often the criterion is established as 10^5 organisms per g. Figure 14-15 shows a comparison between the Bactometer and conventional plate counts for 257 frozen food specimens. Again detection time is compared with the number of organisms as determined by the standard plate count method. The correlation between the plate count method and detection time is 92.6 per-

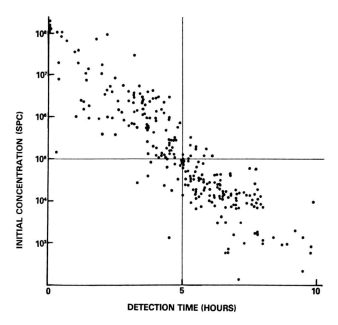

Figure 14-15. Detection times of impedance changes against initial concentration of microorganisms of 257 frozen vegetable samples. The horizontal line at 10^5 organisms per g indicates the demarcation between acceptable and unacceptable levels of contamination. The vertical line at 5 h indicates the cutoff time for the impedance-based test.

cent overall. A detection time of 5 h was found to most correctly resolve frozen food samples into those with greater than 10^5 organisms per g and those with fewer than 10^5 organisms per g. This somewhat longer detection time, compared with urine samples, is due to the somewhat slower generation times found in the populations of microorganisms present in frozen vegetables. Table 14-III shows the results broken down by frozen food category and the corresponding correlation of Bactometer detection times with plate count results (Hardy, 1975). Other foods and beverages can also be handled in a similar fashion. Beer, frozen cream pies, fruit and juices, ground meat, ice cream, packaged gravy mixes, potato salad, soft drinks, and spices have been investigated with the Bactometer.

TABLE 14-III

VARIOUS FROZEN VEGETABLES, THE RESPECTIVE TIMES REQUIRED
FOR IMPEDIMETRIC DETECTION OF BACTERIAL CONCENTRATIONS
ABOVE 10⁵ ORGANISMS PER ML, AND THE PERCENT AGREEMENT
BETWEEN THE IMPEDANCE AND STANDARD PLATE COUNT
METHODS

Frozen Vegetables	Number of Samples	Cutoff Time	Percent Agreement
Green Beans	48	4.9	95.8
Corn	48	4.7	93.7
Peas	47	4.5	95.7
Mixed	41	4.7	95.1
Peas and Carrots	29	5.3	93.1
Others	44	5.3	88.6
All	257	4.9	92.6

Of particular interest is the rapid detection of microorganisms
in milk. A typical impedance curve is shown in Figure 14-16. In
these experiments, the reference well does not contain milk and
the sample well does, thus there is considerable drift noted initially because of the presence of colloidal particles in the sample

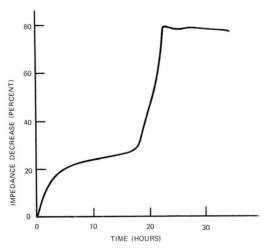

Figure 14-16. Impedance changes resulting from milk bacteria growing in
milk incubated at 25°C and monitored at 400 Hz.

well which are not present in the reference well. This pro-
nounced drift is initially linear, but later decreases in rate and
thus is distingiushable from the accelerating impedance change
due to microorganisms, noted at about 18 h. A rapid assessment
of milk microorganisms, especially psychotrophs, might provide
better data for shelf life predictions — normally a procedure
which requires up to ten days.

Filamentous Fungi

Microorganisms other than bacteria can be detected and
enumerated, e.g. filamentous fungi. Figure 14-17 shows the result
of incubating three different concentrations of *Aspergillus niger*
spores in glucose-peptone-yeast extract broth (GPYE). As with
other organisms, the larger concentration produces the first
initial impedance change, followed by the subsequent lower
dilutions. Of particular interest in this figure is the biphasic

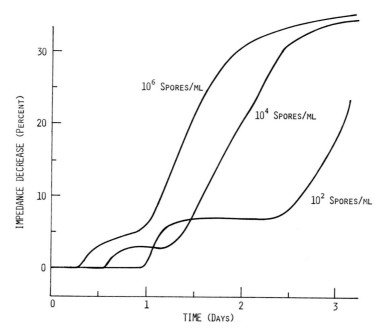

Figure 14-17. Impedance changes resulting from dilutions of *Aspergillus niger* growing in GPYE broth.

nature of all three curves. The initial impedance decrease of from 3 to 6 percent was followed much later by a substantial decrease of nearly 30 percent. Visual observations indicated that the initial impedance decrease was associated with mycelial formation, and that the second impedance decrease had its onset coincident with observable spore formation.

The growth of filamentous fungi is exceptionally difficult to follow with optical measurements, and thus current practice is the tedious gravimetric procedure of dry weight measurements. Figure 14-18 shows the correlation of impedance changes with dry weights of replicate samples of the filamentous fungus *Aspergillus*

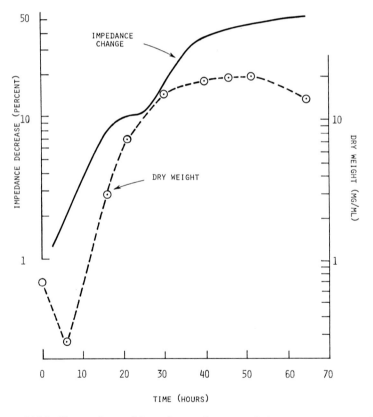

Figure 14-18. Comparison of impedance change and dry weight for *A. niger* growing in GPYE broth.

niger, taken at various times. The initial inoculum for these studies was 10^6 spores per ml. From the graph one can observe that there is a general impedance decrease associated with increasing dry weight. Although the correlation of impedance changes with dry weight is rather poor, the impedance curve provides the additional information that sporulation has occurred, and thus it would seem very likely that continuous monitoring of molds might prove to be valuable in determining optimum growth conditions and the effects of specific nutrients or inhibitors on a laboratory scale, and in monitoring growth characteristics directly in fermentation tanks.

Yeasts

The detection of *Saccharomyces cerevisiae* is depicted in Figure 14-19. Again, impedance decrease versus time is presented for three concentrations of organisms, growing in GPYE broth. Note that, in these examples, the impedance is seen to increase with time. This is depicted by a descending curve, as our convention is to represent impedance decrease by an upward curve. This contrasts with the impedance changes noted for other organisms.

Further light is shed on this phenomenon by Figure 14-20,

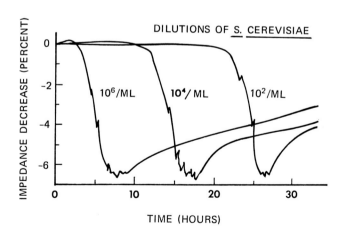

Figure 14-19. Impedance changes resulting from dilutions of *S. cerevisiae* growing in GPYE broth.

obtained for yeasts growing in two media differing only in their glucose concentration. The data suggests that, when the impedance increases, the organisms are more effective in sequestering into their cytoplasm ions from the medium than they were in creating ion pairs from unionized nutrients, thus altering the measured resistive component. On the other hand, changes in medium concentration could also alter the absorption on the electrodes and thus affect the capacitive component. Multiple frequency data, unfortunately not yet available, might clarify this situation.

From a practical viewpoint, impedance measurements are valuable for use with yeasts primarily because of the automated aspects which insure prompt reporting of detection the moment it occurs, rather than detection based on an arbitrary schedule of visual examination of plates by the technologist performing the tests. With the thresholds obtainable with current media (about

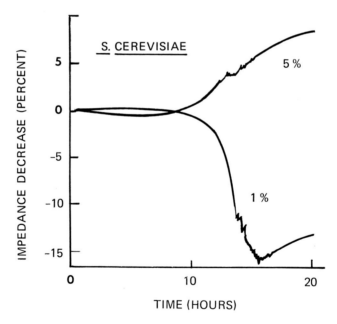

Figure 14-20. Impedance changes resulting from *S. cerevisiae* growing in GPYE broth with 1% and 5% glucose.

1 to 2 × 10⁶ cells per ml), eight to ten yeast cells are detected in about 22 h. This is approximately the same detection time found when colony formation on plates is used. Modification of media may either reduce the threshold or decrease generation times, both of which would shorten detection times considerably.

Anaerobic Bacteria

There is no obvious reason why anaerobic organisms should not give results similar to those obtained with aerobic organisms in impedance measuring instruments, provided that strict anaerobiosis can be readily maintained. Many anaerobic bacteria have rapid generation times and display strong metabolic activities, as evinced, for example, by the proteolytic action of *Clostridia* species.

Hadley and Senyk (1975) have described the detection of anaerobic organisms in blood cultures using the blood culture bottle discussed previously. The medium used, although an excellent general purpose growth medium, was not an optimum medium for anaerobic work and did not use prereduced ingredients. Air was excluded at the time of autoclaving by sealing the bottle while it was still hot and relying upon the small head space gas being mainly water vapor. This gives an environment which is, at best, one of reduced oxygen tension. Hadley reports that the Eh of the medium so prepared is about −110 mV. In this environment, the bottom of the bottle has the lowest Eh and most of the anaerobes isolated have tended to flourish there. In spite of these less than ideal conditions, *Bacteriodes* species and *Clostridium* species have been recovered.

Since impedance measurements can readily be made in a variety of containers and chambers, measurements under more stringent anaerobic conditions should be feasible. Figure 14-21 shows the impedance changes caused by *Bacterioides fragilis* growing in Schaedler medium under strict anaerobic conditions. Note that the response is very similar to that seen with aerobic bacteria.

In general, one would predict that growing anaerobic bacteria under stringent anaerobic conditions would lead to greater metabolic activity and shorter generation times, which would be reflected in faster and greater impedance changes.

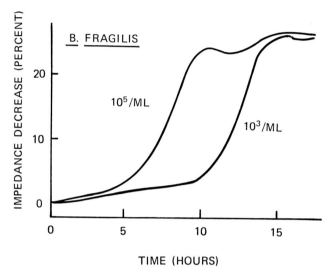

Figure 14-21. Impedance changes resulting from dilutions of *Bacteroides fragilis* (10^5/ml and 10^3/ml inoculated) growing anaerobically in Schaedler broth.

Mycoplasma

The last example of the use of impedance measurement for detection is the application of the Bactometer for the detection of various species of mycoplasma. Mischak (1975) has been successful in detecting a wide range of mycoplasma, including *Ureaplasma urealyticum, Acholeplasma laidlawii, Mycoplasma fermentans, M. pneumoniae, M. pulmonis,* and *M. hominis.* Mycoplasma are ordinarily most difficult to detect and identify. They are fastidious and fragile organisms: The absence of a rigid cell wall subjects them to insults from surfactants well tolerated by their brethren with cell walls and their minute size dictates that even their typical "fried egg" colonies on agar surfaces are best searched for using a hand lens. Their role as the cause of atypical pneumonia in man and of a number of respiratory diseases in cattle and poultry has been described by Hayflick (1972). Considerable interest has been generated recently in investigating their possible role as etiologic agents or coagents in nongonococcal urethritis and certain cases of arthritis.

The upper graph in Figure 14-22 shows *Ureaplasma urealyticum* in three different concentrations growing in urea medium (Masover et al., 1974). The number of organisms is expressed in color-forming units which is an easier measurement to make than the counting of colonies. The impedance change associated with 10^5, 10^4, and 10^3 color-forming units is depicted. The lower

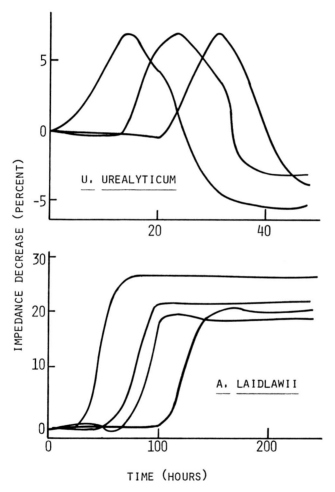

Figure 14-22. Impedance changes resulting from dilutions of *Acholeplasma laidlawii* growing in PPLO broth and *Ureaplasma urealyticum* growing in urea medium.

graph shows *Acholeplasma laidlawii* in the following concentrations: 2×10^5, 10^4, 10^2 and 1 cells per ml. The generation times derived from these curves agree well with the generation times determined by color formation and colony counts.

The growth of mycoplasma has been shown by Razin et al. (1970) to be inhibited by specific antisera. Taking advantage of this phenomenon, Mischak et al. (1975) grew *Ureaplasma urealyticum* and *Acholeplasma laidlawii* in the presence and absence of traces of the specific aintisera appropriate for each organism. Results are depicted in Figure 14-23. The additions of normal serum or of nonhomologous antiserum were without effect.

It should be noted that the impedance response was both delayed in time and diminished in magnitude, suggesting that specific inhibition occurred. We thus have at once a potential detection and identification scheme which is reasonably rapid and certainly simple and convenient.

Multiple Measurements

All of the previous discussion concerns simple detection and enumeration using a single chamber for a single measurement. For mycoplasma work, a multichamber test for identification, that is, one chamber without antiserum and other chambers each with a different antiserum, is suggested. Other multiple measurements include antibiotic susceptibilities, characterization and identification of microorganisms, and bioassays.

Antibiotic Susceptibilities

There is increasing interest in, and acceptance of, rapid tests for antibiotic susceptibility. The advantages to the clinician are obvious and need not be dwelt upon here. There is, however, a basic problem with short term studies of antimicrobial susceptibility which involves the nature of some of the resistance mechanisms themselves. In some instances, for example, with the penicillins, resistance may involve the induction of an enzyme penicillinase. The induction process may be rapid or may require a varying amount of time. Thus, under some circumstances the growth of an organism may be initially suppressed by penicillin, but

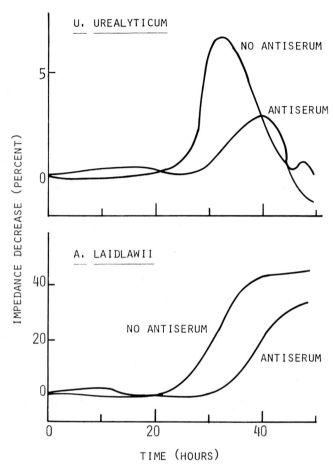

Figure 14-23. The effect of specific antiserum on the growth of *A. laidlawii* in PPLO broth and *U. urealyticum* in urea medium as monitored by impedance measurements.

at a later time it will flourish behind its protective shield of penicillinase. The time at which the organism's response is measured, therefore, becomes important. In general, those organisms showing prompt growth in the presence of a particular antibiotic can safely be considered resistant (at least to that agent in that specific concentration). The converse is not necessarily true. The longer growth is inhibited, however, the greater is the

probability that the organism is susceptible.   The effect of
inoculum size on the result is yet another complicating factor.
Finally, the *in vitro* result may not be as accurate a guide as desired for the final *in vivo* effect.

To a first approximation, it would appear that a 3 to 5 h
incubation gives results which correlate reasonably well with 18
to 24 h diffusion methods or 24 h tube dilution methods (McKie
et al., 1975).   If, during the 3 to 5 h period, continuous growth
measurements can be made, additional confidence can be gained
over any method employing only an initial and final measurement, such as is done with the Pfizer Autobac® or the Technicon®
automatic susceptibility device.   This stems from the knowledge
that the onset of exponential growth, even of a small magnitude
and at the end of the designated time period, has a greater probability of leading to a resistant designation than a similar linear
increase throughout the incubation period.   The most rapid determination of exponential growth is provided by continuous
measurements whether the index of growth is light scattering,
heat production, or impedance change.   Figure 14-24 shows a
typical result when *E. coli* is incubated in Mueller Hinton broth
with penicillin and cephalothin.   The resistance to penicillin is
rapidly established within 90 min.   Confidence in the diagnosis
of susceptibility to cephalothin increases as the experiment pro-
gresses, and the conclusion is supported by simultaneous 24 h
disc diffusion studies.

Of more utility to the clinician is the determination of minimum inhibitory concentrations.   An impedance analog of presently used tube dilution techniques would appear to be very
feasible.

Characterization and Identification

Although little work has been done to date toward applying
impedance measurements to bacterial identification, this field
appears to have great potential for impedance measurements because of the ease with which large numbers of channels can be
sampled.   Figure 14-25 shows curves demonstrating the feasibility
of determining an organism's sugar assimilation patterns from

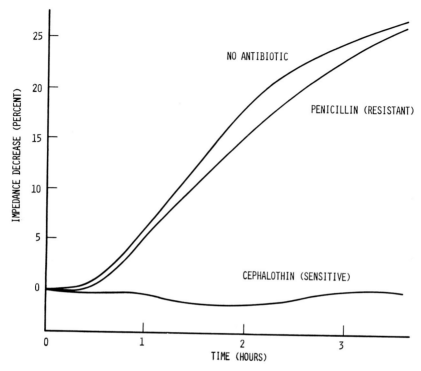

Figure 14-24. The effect of penicillin and cephalothin on the growth of *E. coli* (4 x 10⁷/ml inoculated) in TSB as monitored by impedance measurements.

impedance measurements. This figure shows the impedance changes resulting from 10⁷ cells per ml of *Neisseria gonorrhoeae* inoculated into Thayer Martin broth and Thayer Martin broth with maltose and sucrose substituted for the glucose. The fact that *N. gonorrhoeae* assimilates glucose, but not maltose or sucrose, is readily apparent within 10 h. Some of the same considerations discussed under susceptibility testing also apply here. With most present fermentation tests, the test is not considered negative until at least 18 to 24 h have elapsed. With some organisms even longer times are required. Without continuous measurements it is not easy to rapidly differentiate between an organism with a diminished ability to utilize a particular carbohydrate

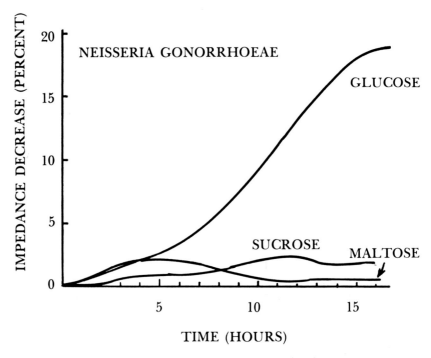

Figure 14-25. Impedance changes resulting from *Neisseria gonorrhoeae* (10⁷/ ml inoculated) in Thayer Martin Broth and Thayer Martin Broth with its glucose replaced by sucrose and maltose.

substrate and one which has not as yet fully induced the appropriate enzymes necessary to handle the substrate.

An extension of this concept is to use inhibitory media such as antibiotics and dyes, so as to build a repertory of differential environments such that the results of growth in such a set of media can be used to further characterize and even identify an unknown organism. Such a set of media has been proposed by Matsen (Chapter 15) for use in optical growth measuring systems.

Economic Considerations

Impedance measurement methods realize considerable savings in costs for large-scale routine testing. In this case the major saving is the decreased preparation time required for such procedures as standard total plate counts of food products. A single chamber in a disposable module together with its medium replaces the plates, media, and dilution tubes needed for the several dilutions required. These two sets of costs are comparable; however, the preparation time in the former is considerably less. In addition, samples with particulate matter pose some difficulty in plate count interpretation, which is not the case in impedance procedures. There may be greater economic advantages, resulting from reductions in analysis time, than the cost reduction per test. For example, gross food processing contamination may be detectable within an hour, and thus remedial steps can be taken much sooner.

REFERENCES

Allison, J.B., J.A. Anderson, and W.H. Cole: 1938. The method of electrical conductivity in studies on bacterial metabolism. *J Bacteriol, 36:* 571.

Andreev, V.S.: 1974. The Titr-2 semiautomatic conductometric titrator *Biomed Eng, N.Y. 8:*141.

Cady, P.: 1975. Rapid automated bacterial identification by impedance measurement. In *New Approaches to the Identification of Microorganisms,* ed. by C. Hedén and T. Illéni, pp. 73 Wiley, New York.

Carstensen, E.L., H.A. Cox, Jr., W.B. Mercer, and I.A. Natale: 1965. Passive electrical properties of microorganisms. I. Conductivity of *Escherichia coli* and *Micrococcus lysodeikticus. Biophys J, 5:*289.

Carstensen, E.L.: 1967. Passive electrical properties of microorganisms. II. Resistance of the bacterial membrane. *Biophys J, 7:*493.

Carstensen, E.L. and R.E. Marquis. 1968. Passive electrical properties of microorganisms. III. Conductivities of isolated bacterial cell walls. *Biophys J, 8:*536.

Geddes, L.A. and L.E. Baker: 1968. Detection of physiological events by impedance. In *Principals of Applied Biomedical Instrumentation,* pp. 150. Wiley, New York.

Geddes, L.A., L. Baker, A. Moore, and T. Coulter: 1969. Hazards in the use of low frequencies for the measurement of physiological events by impedance. *Med Biol Eng, 7:*289.

Geddes, L.A., C. DaCosta, and G. Wise: 1971. The impedance of stainless steel electrodes. *Med Biol Eng, 9:*511.

Goldschmidt, M.C. and T.G. Wheeler: 1973. Determination of bacterial cell consentrations by bacterial impedence measurements. *Abstr Ann Meet Am Soc Microbiol, E48.*

Hadley, W.K. and G. Senyk: 1975. Early detection of microbial metabolism and growth by measurement of electrical impedance. In *Microbiology*

1975, edited by D. Schlessinger. American Society for Microbiology, Washington, D.C.

Hanss, M. and A. Rey: 1971a. Use of conductometry in the study of enzymic reactions. Butyryl cholinesterase. *Biochim Biophys Acta, 227:* 618.

Hanss, M. and A. Rey: 1971b. Use of conductometry in the study of enzymic reactions. Ureaurease system. *Biochim Biophys Acta, 227:*630.

Hardy, D.: 1975. Rapid detection of frozen food bacteria by automated impedance measurements. Inst. of Food Technologists Annual Meeting, Chicago.

Harvey, R.J., A. Marr, and P. Painter: 1967. Kinetics of growth of individual cells of *Escherichia coli* and *Azotobacter agilis*. *J Bact, 93:*605.

Hayflick, L.: 1972. Mycoplasmas as pathogens. In *Pathogenic Mycoplasma —A Ciba Foundation Symposium,* N.W. Pirie (chairman), pp. 17. Elsevier, Amsterdam.

Kass, E.H.: 1955. Chemotherapeutic and antibiotic drugs in the management of infections of the urinary tract. *Am J Med, 18:*764.

Lawrence, A.J.: 1971. Conductometric enzyme assays. *Eur J Biochem, 18:* 221.

Lawrence, A.J. and G.R. Morres: 1972. Conductometry in enzymes studies. *Eur J Biochem, 24:*538.

Masover, G.K., J.R. Benson, and L. Hayflick: 1974. Growth of T-strain mycoplasmas in medium without added urea: effect of trace amounts of urea and of a urease inhibitor. *J Bact, 117:*765.

McKie, J.E., R. Borovoy, J. Dooley, G. Evanega, G. Mendoza, F. Meyer, M. Moody, D. Packer, J. Praglin, and H. Smith: 1975. Autobac 1—a three hour automated antimicrobial susceptibility system, II microbiological studies. In *Automation in Microbiology and Immunology,* ed. by C. Hedén and T. Illéni, pp. 209, New York.

McPhillips, J. and N. Snow: 1958. Studies on milk with a new type of conductivity cell. *Aust J Dairy Technol, 3:*192.

Mischak, R.P., P. Cady, and S.J. Kraeger: 1975. Detection of mycoplasma by impedance measurements. *Abstr Ann Meet Am Soc Microbiol, Q17.*

Parsons, L.B. and W.S. Sturges. 1926a. The possibility of the conductivity method as applied to studies of bacterial metabolism. *J Bact, 11:*177.

Parsons, L.B. and W.S. Sturges: 1926b. Conductivity as applied to studies of bacterial metabolism. *J Bact, 12:*267.

Parsons, L.B., E.T. Drake, and W.S. Sturges. 1929. A bacteriological conductivity culture cell and some of its applications. *J Am Chem Soc, 51:* 166.

Razin, S., B. Prescott, G. Caldes, W. James, and R. Chanock: 1970. Role of glycolipids and phosphatidylglycerol in the serological activity of *M. pneumoniae. Infect Immun, 1:*408.

Schwan, H.P.: 1963. Determination of biological impedances. In *Physical Techniques in Biological Research,* ed. by W. Nastuk, pp. 323 Acad Pr, New York.

Schwan, H.P.: 1968. Electrode polarization impedance and measurements in biological materials. *Ann NY Acad Sci, 148*:191.

Stacy, R.W.: 1960. *Biological and Medical Electronics.* Chapter 3. McGraw, New York.

Stewart, G.N.: 1899. The changes produced by the growth of bacteria in the molecular concentration and electrical conductivity of culture media. *J Exp Med, 4*:235.

Ur, A.: 1970. Changes in electric impedance of blood during coagulation. *Nature,* London, 226:269.

Ur, A. and D.F.J. Brown: 1974. Rapid detection of bacterial activity using impedance measurements. *Biomed Eng, 9*:18.

Ur, A. and D.F.J. Brown: 1975a. Imepedance monitoring of bacterial activity. *J Med Microbiol, 8*:19.

Ur, A. and D.F.J. Brown: 1975b. Monitoring of bacterial activity by impedance measurements. In *New Approaches to the Identification of Microorganisms,* ed. by C. Hedén and T. Illéni. Wiley, New York.

VanderTouw, F., J. Briede, and M. Mandel: 1973. Electrical permittivity of alfalfa mosaic virus in aqueous solutions. *Biopolymers, 12*:111.

Warburg, E.: 1899. Uber das verhalten sogenannter unpolarisirbarer Electroden gegen Wechselstrom. *Ann Phys Chem, 67*:493.

Warburg, E.: 1901. Uber die Polarizations capacitat des platins. *Ann Phys, 6*:125.

Wheeler, T.G. and M.C. Goldschmidt: 1975. Determination of bacterial cell concentrations by electrical measurements. *J Clin Microbiol, 1*:25.

Chapter 15

RAPID AUTOMATED BACTERIAL IDENTIFICATION WITH COMPUTERIZED PROGRAMMING OF AUGMENTED AUTOBAC 1 RESULTS*

J. M. MATSEN, B. H. SIELAFF, AND G. E. BUCK

Abstract

The Autobac 1, an automated instrument for antimicrobial susceptibility testing, can be used to provide bacterial identification in the same 3 to 5 h time frame employed for susceptibility testing. The identification system involves the use of antibiotics normally used in antimicrobial susceptibility tests, as well as other chemicals that selectively inhibit growth. A computer program has been developed which analyzes these results by means of the quadratic discriminant function statistical technique. Accuracy of identification for various taxonomic groupings of bacteria is 95 percent or greater, as compared to standard biochemical methods.

Economic, professional, and patient care pressures have combined to motivate development of systems which will produce more rapid results from the clinical microbiology laboratory. Because machines and computerization have served so well to assist science and industry in obtaining rapid answers to problems presented in many and varied spheres of activity, it was inevitable that automated methods for providing rapid answers in clinical microbiology would also be explored. In 1974 a system was reported by Praglin et al. and McKie et al. which carried the name of Autobac 1. This is an automated system for measuring antimicrobial susceptibilities of bacterial organisms within a three-

*This study was funded in part by grants from the Pfizer Diagnostics Division. The excellent technical assistance of Becky Boshard and secretarial assistance of Glenna Bytendorp is gratefully acknowledged.

Figure 15-1. Autobac 1 system components. Light scattering photometer, upper right; cuvette, lower center; disc dispenser, lower left; and incubator shaker, upper left.

to five-hour time period. The system is basically comprised of four separate components (Fig. 15-1), each carefully designed to perform a specific function leading to the end result. The first and central component of the system is a light scattering photometer (upper right), the second a cuvette comprised of thirteen individual chambers (lower center), the third a dispenser which delivers antimicrobic discs to the individual chambers of the cuvette (lower left), and the fourth a combination incubator and shaker (upper left). The antimicrobic discs are the source of the antimicrobial agents used for determining bacterial susceptibility; as will be described later, they are also used as a source of other growth no-growth chemicals used in the differentiation of organism groupings. The interrelation of these four components effectively produces results on antimicrobial susceptibility and data which can be used for the identification of organisms.

Because the results generated in a collaborative study involving seven different clinical centers (Thornsberry et al., 1975) indicated that the Autobac 1 susceptibility system compared favorably with the standardized disc diffusion method (Bauer, 1966; Natl. Comm. Clin. Lab. Std., 1974) and the standardized agar dilution technique (Ericsson and Sherris, 1971), we felt that this system had sufficient promise to warrant a major effort to adapt it for the identification of organisms. The results of these efforts are reported in this paper.

To understand the basic premise upon which this identification system is built, it is first necessary to understand in some detail the workings of the system. A more detailed description is provided in the report of the collaborative study (Thornsberry et al., 1975).

Plastic cuvettes, stored in plastic bags, are removed from their containers and the white, flexible plastic closures are stripped back from their tops. Each cuvette is then inserted into the disc dispenser and the dispenser activated, releasing selected antimicrobic discs into each of the chambers of the cuvette (Fig. 15-2). Only one disc is delivered to each chamber and the white strip closure is then replaced. The material contained in each disc is previously selected for either antimicrobial susceptibility or for ability to differentiate between bacterial species.

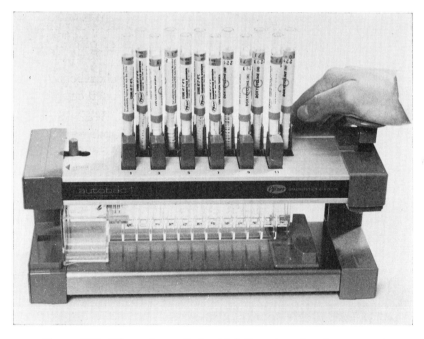

Figure 15-2. Dispensing antimicrobial discus into Autobac cuvette.

The next step is to select a bacterial colony from the primary isolation plate and to inoculate cells of it into a tube of normal saline. This tube is then inserted into the light scattering photometer (Fig. 15-3) and saline added if necessary to standardize the inocula. When the meter needle registers in the center portion of the marked gauge, standardization has occurred. Two ml of the

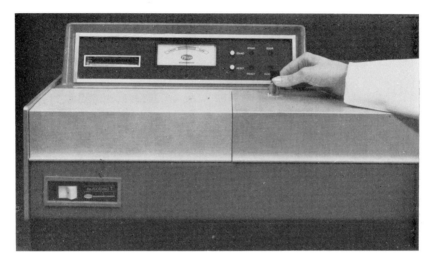

Figure 15-3. Standardizing the saline inoculum suspension.

standardized inoculum are then pipetted into a tube of eugonic broth. While holding the cuvette vertically, the eugonic broth tube is screwed into the cuvette (Fig. 15-4) and the cuvette then inverted on a level surface (Fig. 15-5) so that all of the broth is distributed to a holding chamber situated at one end of the cuvette. Further manipulations of the cuvette are made to distribute the broth evenly along the length of the cuvette (Fig. 15-6) and to deliver 1.5 ml aliquots of the broth into the individual chambers of the cuvette (Fig. 15-7), which were previously armed with the antimicrobic discs.

The cuvette is then placed in the incubator shaker (Fig. 15-8) and rotated at 220 rpm at 36°C for approximately 3 h. Next, the cuvette is placed on a holding bar or carriage in the light scatter-

Figure 15-4. Attaching the inoculated broth tube to the cuvette.

Figure 15-5. Filling the cuvette reservoir.

Figure 15-6. Distributing the inoculated broth evenly along the length of the cuvette.

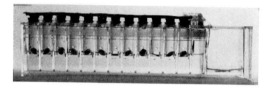

Figure 15-7. Delivering equal volumes of broth to each of the chambers.

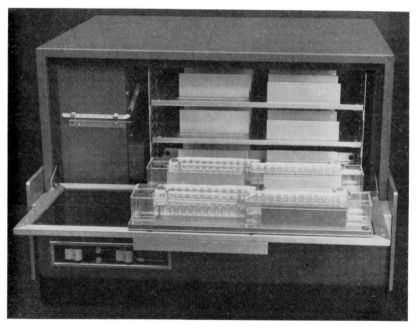

Figure 15-8. Placing the cuvette in the incubator/shaker.

ing photometer (Fig. 15-9), and the photometer lid closed. A card is inserted to record the results, and the machine then begins its computation. Sufficient growth must be obtained in the control chamber or the photometer will automatically reject the test. When this occurs, the cuvette is returned to the incubator and incubated for an additional period. If there is sufficient growth in the control chamber, then a light scatter index is calculated by means of a mini-computer within the photometer housing. This index is a numerical value between 0 and 1 with 0.01 subdivisions, and based on it, an interpretation of susceptible, intermediate, or resistant is calculated for each antimicrobial agent.

The light scattering index for each antimicrobial agent used is utilized in a multivariate analysis procedure called the quadratic discriminant function. This function is based on the assumption that members of each of the bacterial groups tend to follow a

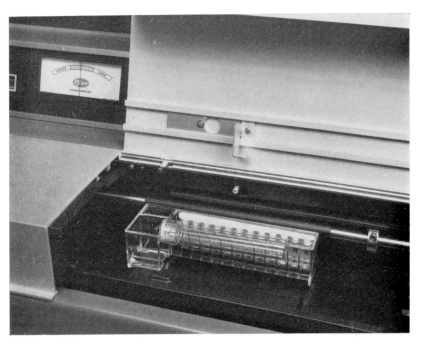

Figure 15-9. The cuvette in place in the Autobac photometer ready to have results determined.

characteristic normal distribution with each of the antimicrobial compounds tested (Sielaff et al., 1976). If two groups of bacteria happen to have overlapping distribution curves, then a point of equal probability is created in the overlap region. This point of equal probability can be used as a boundary for classification. If there is considerable overlap, then even a sophisticated computer technique cannot separate groups of bacteria on the bases of one variable. As one adds variables, a greater likelihood of separation occurs. If the groups have widely differing values with each of several antimicrobial agents, then misclassification is minimized, and identification is rather straight forward and easily accomplished.

It is important, therefore, to use compounds which discriminate widely on the basis of light scattering index values or to have at least one compound which provides a clear-cut distinction.

Initially, the authors' efforts were concerned with only those agents which have been used therapeutically in humans against bacterial organisms. Gram-negative organisms were chosen for the work because of the generally greater difficulty in separating them into the various clinically significant species. Table 15-I lists according to genus or species the 481 cultures tested with the first antimicrobial profile. The antimicrobial compounds employed in the profile are shown in Table 15-II. The matrix, or data base, utilized in the quadratic discriminant function computer program was generated by feeding to the computer the light scattering index values from known organisms within these bacterial group-

TABLE 15-I

481 ORGANISMS TESTED WITH PROFILE I AND II

		Other *Proteus*	
Acinetobacter	35	*morganii*	19
Citrobacter	50	*rettgeri*	17
Enterobacter	48	*vulgaris*	15
Escherichia coli	75	*Pesudomonas*	
Klebsiella	59	*aeruginosa*	35
Proteus mirabilis	49	*fluorescens*	15
Serratia marcescens	52	*maltophilia*	12

TOTAL: 481 Strains

TABLE 15-II

ANTIMICROBIC PROFILE I (18 AGENTS)

	μg		μg
Ampicillin	3.6	Kanamycin	5
Bacitracin	18 (u)*	Methenamine	1
Carbenicillin	50	Naladixic Acid	5
"	120	Neomycin	5
Cephalothin	15	Nitrofurantoin	15
Colistin Sulfate	2	Novobiocin	30
"	13	Polymyxin B	50 (u)
Erythromycin	15	Streptomycin	10
Furizolidone	100	Tetracycline	0.5

*(u) = Units, not μg.

ings. An accuracy of 97.3 percent was achieved for this eighteen-agent profile (Sielaff et al., 1976).

Problems, however, exist with respect to the size of the profile, the limited number of species involved in the data base, and the possibility that resistance to some of the antimicrobial agents might have been developed in organisms found in human patients, animals, animal feeds, or even in the environment generally.

The next step in the study was an attempt to decrease the antimicrobic profile, in order to reduce the number of variables to a minimum without sacrificing percentage agreement. In this particular endeavor, various combinations of the eighteen agents listed in Table 15-II were used. Table 15-III shows one particular grouping of fourteen agents. When this was tested for the 481 organisms previously described, there was a loss in agreement of

TABLE 15-III

ANTIMICROBIC PROFILE II (14 AGENTS)

	μg		μg
Ampicillin	3.6	Erythromycin	15
Bacitracin	18 (u)*	Kanamycin	5
Carbenicillin	50	Methenamine	1
"	120	Neomycin	5
Cephalothin	15	Nitrofurantoin	15
Colistin SO$_4$	2	Novobiocin	30
"	13	Tetracycline	0.5

*(u) = Units, not μg.

less than 2 percent. As the number of agents was reduced below fourteen, however, the percentage agreement dropped rapidly. Therefore, fourteen antimicrobials appeared to be the smallest group which still gave an acceptable level of agreement (Sielaff et al., 1976). The group contained only agents that are commonly used for therapeutic purposes.

Next, attempts were made to develop a repertoire of compounds which have antimicrobial action, but which are not classified as antibiotic or therapeutic agents. In this investigation, Buck, Boshard, and Matsen reviewed over 600 compounds for known or potential antibacterial activity. Approximately 15 percent (104 compounds) were selected for screening to determine their differential selectivity against twenty different species (Table 15-IV), including the majority of the Enterobacteriaceae, as well as commonly isolated members of the nonfermenting gram-negative groups. Extensive tests were made to determine the minimum levels at which antibacterial activity occurred for the organisms studied; over 10,000 sets of tests were performed. Those compounds which showed capacity for differentiating bacterial groups were considered for further analysis in the computer system.

Initially, a small number of these compounds were used in con-

TABLE 15-IV

*BACTERIAL GROUPS USED IN SCREENING COMPOUNDS
FOR THE ABILITY TO SEPARATE BACTERIA
BASED ON GROWTH INHIBITION*

(1) *Escherichia coli*	(11) *Serratia*
(2) *Citrobacter freundii*	(12) *Klebsiella*
(3) *Citrobacter diversus*	(13) *Providencia*
(4) *Enterobacter*	(14) *Acinetobacter calcoaceticus* var. *antiratus*
(5) *Salmonella*	(15) *Acinetobacter calcoaceticus* var. *lwoffi*
(6) *Shigella*	(16) *Edwardsiella*
(7) *Proteus mirabilis*	(17) *Arizona*
(8) *Proteus morganii*	(18) *Pseudomonas aeruginosa*
(9) *Proteus rettgeri*	(19) *Pseudomonas maltophilia*
(10) *Proteus vulgaris*	(20) Other *Pseudomonas* species

junction with agents previously investigated. Twelve compounds (enough to fill the test chambers of one cuvette) were utilized in this part of the investigation, and included five from the new group (Table 15-V). Fourteen of the original twenty bacterial groups, with twenty-four or twenty-five isolates per group (Table 15-VI), were then tested against these compounds. Purposely included again were organisms which are difficult to separate such as the *Escherichia, Citrobacter, Enterobacter,* and *Proteus* general (Buck et al., 1975).

TABLE 15-V

ANTIMICROBIC PROFILE III

	µg		µg
*Brilliant Green	3	*Methenamine	1
Carbenicillin	120	Nitrofurantoin	15
Cephalothin	15	Polymyxin B	12.5 (u)†
*Chlorhexidine	4	Tetracycline	0.5
		*Trihydroxyaceto-	90
*Cycloserine	30	phenone	
Doxycycline	0.5	Streptomycin	10

*Agents not used in either Profiles I or II.
†(u) = Units, not µg.

An accuracy of 97.1 percent was achieved in the computer program, using the calculated rules. The poorest results were obtained with *Escherichia coli,* where only twenty of the twenty-four strains tested were correctly identified. Three *Citrobacter freundii* cultures were misidentified as was one *Enterobacter* strain and two *Proteus vulgaris* strains. The problem with *Escherichia*

TABLE 15-VI

346 ORGANISMS TESTED WITH PROFILE III

Acinetobacter (Var. *antiratus*)	25	*Proteus morganii*	25
Citrobacter diversus	25	*Proteus rettgeri*	25
Citrobacter freundii	24	*Proteus vulgaris*	25
Enterobacter	25	*Pseudomonas aeruginosa*	24
Escherichia coli	24	*Salmonella* sp.	25
Klebsiella pneumoniae	25	*Serratia marcescens*	24
Proteus mirabilis	25	*Shigella*	25

coli is in separating it from *Shigella* species. However, the authors' screening studies showed that agents do exist which will separate these genera.

The studies objective, already stated, has been successfully accomplished. Considerable time over the past two years has been spent in identifying differentiating compounds and in creating a computer program which will successfully employ the machine-generated values. To prevent dilution of effort in the feasibility studies, it was necessary to limit the number of microbial groups studied. The focus was on the gram-negative bacteria for several reasons. First, the gram stain is a fairly quick and easy method of differentiating this group from other bacterial groups. Second, the members of this group are generally the most difficult to identify; they provide the greatest challenge for the proposed identification system. Third, gram-negative bacteria comprise the majority of microbial identifications being done currently in clinical laboratories.

As mentioned, special attention is being paid to the problem of developed resistance to various antimicrobial compounds. As shown in the authors' third profile (Table 15-V), compounds are being used which are not used therapeutically and for which, therefore, resistance has not been developed. Even greater reliance on such compounds is planned. To inform interested laboratories on resistance problems and on development of new agents, the authors propose that a national data base be generated by the manufacturer of the Autobac system. For this purpose, the authors have been utilizing a computer terminal in the research laboratory into which data is fed, stored on a magnetic tape cassette, and sent quickly by telephone to a central computing facility. There are several ways in which such a system can be utilized to quickly inform laboratories on changes in organism susceptibility profiles and on improved profiles.

This chapter describes a limited approach to the identification necessary in the clinical microbiology laboratory to handle the many clinical specimens sent to it for identification. Work is underway to expand the profile to include all significant clinical bacteria; in this connection, it is noteworthy that preliminary work has shown that identification of gram-positive organisms is also feasible by this system.

As indicated, further work is being done on the chemical compounds screened for their activity against various gram-negative bacteria. The authors now have a repertoire of forty compounds which show promise. The goal is to select ten or twelve to supplement information obtained from routinely used antibiotic agents. The ultimate finesse of the system may not occur for some time, but it seems feasible at this point to begin clinical studies in a laboratory setting to confirm, in parallel tests, these research results. Such studies are currently planned, and will be undertaken as soon as discs can be uniformly manufactured. The desire is to use the same type and quality of disc that would be commercially available subsequently.

REFERENCES

Bauer, A.W., W.M.M. Kirby, J.C. Sherris, and M. Turck: 1966. Antibiotic susceptibility by a standardized single disc method. *Am J Clin Path, 45:* 493.

Buck, G.E., B.H. Sielaff, and J.M. Matsen: 1975. Automated, rapid identification of bacteria by computer analysis of growth inhibition patterns with Autobac 1. Fifteenth Interscience Conference on Antimicrobial Agents and Chemotherapy. Abstract No. 365, Washington, D.C., September 24-26.

Ericsson, J. and J.C. Sherris: 1971. Antibiotic susceptibility testing. Report of an international collaborative study. *Acta Pathol Microbiol Scand Suppl 217.*

McKie, J.E., Jr., R.J. Borovoy, J.F. Dooley, G.R. Evanega, G. Mendoza, F. Meyer, M. Moody, D.E., Packer, J. Praglin, and H. Smith: 1974. Autobac 1—A 3-hour automated, antimicrobial susceptibility system: II Microbiological Studies. In *Automation in Microbiology and Immunology,* ed. by C. Hedén and T. Illéni. Wiley, New York.

National Committee for Clinical Laboratory Standards Subcommittee on Antimicrobial Susceptibility Testing: 1974. Tentative performance standards for antimicrobial susceptibility tests as used in clinical laboratories, p.p. 138-155. In *Current Techniques for Antibiotic Susceptibility Testing,* ed. by A. Balows. Thomas, Springfield.

Praglin, J., A.C. Curtis, D.K. Longhenry, and J.E. McKie, Jr.: 1974. Autobac 1—A 3-hour automated antimicrobial susceptibility system: I. System description. In *Automation in Microbiology and Immunology,* ed. by C. Hedén and T. Illéni. Wiley, New York.

Sielaff, B.H., E.A. Johnson, and J.M. Matsen: (In press). Computer assisted bacterial identification utilizing antimicrobial susceptibility profiles generated by Autobac 1. *J Clin Microbiol.*

Thornsberry, C., T.L. Gavan, J.C. Sherris, A. Balows, J.M. Matsen, L.D. Sabath, F. Schoenknecht, L.D. Thrupp, and J.A. Washington, II: 1975. Laboratory evaluation of a rapid, automated susceptibility testing system: Report of a collaborative study. *Antimicrob Ag Chemother, 7:*466.

Chapter 16

RAPID ANTIMICROBIAL SUSCEPTIBILITY TESTING OF ANAEROBIC BACTERIA WITH AUTOBAC 1*

D. W. LAMBE, JR., A. CURTISS, W. W. LASLIE, J. McKIE, AND J. SEO

Abstract

A new automated method, based on the Autobac 1 system, was developed for rapidly determining the antimicrobial susceptibility of certain anaerobic bacteria. The modifications of the Autobac 1 system included the selection of a prereduced eugonic medium for growth of anaerobes, an initial inoculum of 1 to 2 x 10^7 colony forming units per ml, an Autobac growth index of 0.4, and an elution disk with a penicillin G content of 0.8 unit. To determine the optimal penicillin G content for the disks an interpretive scheme for penicillin G was established by statistical analysis. Final interpretive readings of susceptible, indeterminate, or resistant are printed by the Autobac 1 machine.

INTRODUCTION

Within the past year, instrument systems have become available for rapidly determining the antimicrobial susceptibility of aerobic bacteria. One such system is the Autobac 1 instrument which determines antimicrobial susceptibility of aerobic bacteria within three to five hours (Thornsberry et al., 1975), as compared to the Kirby-Bauer disk method (Bauer et al., 1966) which requires 18 to 24 h for test results. The 15 to 21 h saving in time by the Autobac machine is an obvious benefit in treatment of the patient. This study is the first to adapt the Autobac system to anaerobic bacteria.

*We very much appreciate the expert technical assistance of Kay N. Bishop, Catherine Tyler, and David A. Lord.

Previously, a conventional disk test for anaerobes was developed in the author's laboratory for tetracycline (Overman et al., 1974). Sutter et al. (1972) and Wilkins et al. (1972) have also described a disk test method. However, the disk method for anaerobes usually requires a minimum of two days between the appearance of the colony on the primary isolation plate and the reporting of antimicrobial susceptibility test results to the clinician. The delay in reporting results is a decided disadvantage in treating patients with anaerobic infections. A more rapid test is needed to aid the physician in the early selection of the proper antimicrobial agent to combat anaerobic infection, especially life threatening infections such as anaerobic septicemia. Furthermore, routine antibiotic testing of anaerobes is becoming more important with the appearance of drug resistance among anaerobic strains, for example, tetracycline-resistant and, more recently clindamycin-resistant strains of *Bacteroides fragilis.*

This investigation was initiated to determine if the automated technology of the Autobac 1 could be adapted to provide rapid antimicrobial susceptibility testing of anaerobic bacteria. The authors will describe an Autobac system which can be used to determine the antibiotic susceptibility of rapidly growing anaerobes to penicillin G. For the purpose of this study, rapidly growing anaerobes are those that will grow in eugonic broth between 3 and 24 h. Strains requiring more than 24 h were considered slow growing and because of this were not considered appropriate for this rapid automated test.

MATERIALS AND METHODS

The study included selection of the proper test inoculum for the Autobac, selection of a prereduced broth which supported growth of most anaerobic clinical isolates, determination of the proper growth index of anaerobic bacteria for the Autobac instrument, and development of an Autobac method using penicillin G for the model system.

Strains

Clinical isolates (95 strains) used included: *Bacteriodes fragilis* subsp. *distasonis* (10 strains); *B. fragilis* subsp. *fragilis* (20 strains) ; *B. fragilis* subsp. *thetaiotaomicron* (10 strains) ; *B. fragilis* subsp. *vulgatus* (10 strains) ; *Fusobacterium necrophorum* (2 strains) ; *Fusobacterium varium* (2 strains) ; *Actinomyces israelii* (1 strain) ; *Actinomyces viscosus* (1 strain) ; *Actinomyces naesulundii* (1 strain) ; *Bifidobacterium species* (2 strains); *Eubacterium lentum* (1 strain) ; *Eubacterium limosum* (1 strain) ; *Peptoccus magnus* (9 strains) ; *Peptococcus prevotii* (3 strains) ; *Peptostreptococcus anaerobius* (10 strains) ; *Streptococcus constellatus* (3 strains) ; *Streptococcus intermedius* (8 strains) ; and *Streptococcus morbilloru* (1 strain) .

Growth Medium

In the Autobac system for aerobes, eugonic broth is used as a growth medium, and an initial inoculum of approximately 1.5 to 3×10^6 colony forming units (CFU) per ml is employed. The aerobes have grown sufficiently when a growth index of 0.9 is reached. When the organism has not grown to this level, the machine will not read the light-scattering index and the susceptibility of the organism is not determined.

Thus, using 0.9 as an indicator of growth, studies were performed with sixty-two representative strains of anaerobes to determine the ability of three prereduced broths to support growth. The three broths were eugonic broth (EB) , Schaedler's broth (SB) (BBL) , and brain heart infusion broth (BHIB) (Difco). Prereduced phosphate buffered physiological saline (PBS) was used as a diluent. The broths and the PBS were prereduced to an Eh of -200 to -250 mV; all fluids contained the redox indicator resazurin. Vitamin K (0.00005%) (Sigma) and hemin (0.0005%) (Sigma) were added as growth factors.

Atmosphere

All tests were performed with the Autobac incubator/shaker in an anaerobic atmosphere of 85 percent nitrogen, 5 percent carbon dioxide, and 10 percent hydrogen. Gas tight cuvettes from which all oxygen was removed were used throughout the study; thus, determination of growth in the cuvette was performed with the photometer in an aerobic environment.

Test Protocol

In the test protocol, anaerobes were grown for 24 to 48 h on blood agar plates, described previously by Overman et al. (1974). Sufficient bacteria were removed from the plates to give an initial inoculum of 1.5 to 3 x 10^8 colony forming units (CFU) per ml in prereduced PBS. A 1:10 dilution of the inoculum was prepared by adding 2 ml of the PBS to 18 ml of prereduced eugonic broth. The inoculated broth tube was then screwed onto the cuvette and the broth evenly distributed throughout the thirteen-chamber cuvette. Chamber one was a control chamber containing no antibiotic. Penicillin G disks ranging in content from 0.1 to 4.0 units, prepared in the authors' laboratory, were placed in the remaining twelve chambers. The cuvettes were incubated anaerobicallly at 37°C in an incubator shaker. Photometer readings of growth were made at 3, 4, 5, 6, and 24 h. Duplicate runs were performed on different days.

RESULTS AND DISCUSSION

Inoculum and Growth Medium

The effects of inoculum and growth medium are shown in Table 16-I. The best growth occurred with an inoculum of 10^7 CFU per ml in eugonic broth. Therefore, this inoculum and this growth medium were used in subsequent studies. Because of the slower growth rate of many of the species and their inability to reach a growth index of 0.9, it was necessary to determine the minimum level of control signal necessary for a reliable reading of growth. To accomplish this, the circuitry of the Autobac was adjusted to permit readings to be taken at any level of growth. A representative number of strains of each species was used to deter-

TABLE 16-I

GROWTH OF 62 ANAEROBIC BACTERIA IN 3 BROTHS USING 0.9 AS
THE GROWTH INDEX (PER CENT POSITIVE IN 6 H)

Inoculum (organisms/ml)	Eugonic Broth	Schaedler's Broth	Brain Heart Infusion Broth
1-2 x 16^6	34	31	16
1-2 x 10^7	47	42	40

mine the effect of growth on susceptibility of the species to penicillin G. A growth index of 0.4 was required before an accurate
determination of susceptibility could be determined. Only 47 percent of the strains grew under the test conditions even when the
growth index was lowered from 0.9 to 0.4.

Interpretive Criteria

The philosophy which guided the instrumental approach used
in this study was to incorporate the principles of susceptibility testing already well established in the Autobac test system. Thus,
interpretive schemes were developed through the correlation of
minimal inhibitary concentrations (MIC) values obtained by the
agar dilution method, with zone size measurements, by the high
content disk diffusion method. For the disk diffusion test and
MIC determinations, Mueller-Hinton agar supplemented with
sheep blood, vitamin K, and hemin was used. A previous report
from the authors' laboratory (Overman et al., 1974) revealed that
this medium is reliable for disk testing of anaerobes and gives
results similar to those of the Kirby-Bauer antibiotic susceptibility
test for aerobes.

Figure 16-1 shows the distribution of the zone diameters of
eighty-seven strains of anaerobes to penicillin G. Because of the
bimodal distribution of strains, interpretation of the disk test was
based on analysis of each of the two populations. Using this type
of population analysis, the interpretive criteria for penicillin G
were as follows: An anaerobe is resistant to penicillin G if the
zone around a 10 unit penicillin G disk is 8 mm or less and susceptible if the zone is 21 mm or greater. Susceptibility of the
anaerobe is indeterminate if its zone of inhibition is between 9 and
20 mm.

Once interpretive criteria for penicillin G were established, it
was necessary to determine the proper elution disk content which
would give satisfactory correlation between Autobac and disk test
results. The correlation of Autobac susceptibility results with
zone size is shown in Table 16-II. The approach taken was to
find a disk content which would yield susceptibility results closely
correlating with the disk diffusion test. The best overall correla-

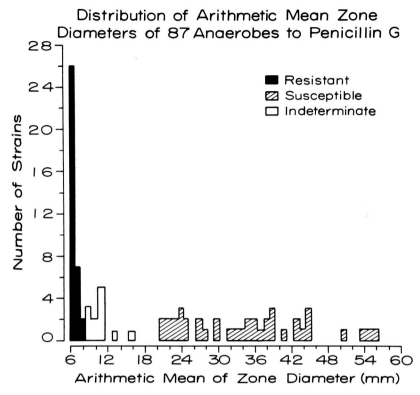

Figure 16-1. Distribution of arithmetic mean zone diameters of eighty-seven anaerobes to penicillin G.

tion, 83.2 percent, occurred with the 0.8 unit disk. The 16.8 percent discrepancy was divided on the basis of the results obtained as follows: 1.0 percent of the strains was susceptible by Autobac but resistant by disk diffusion (very major discrepancies); 2.1 percent of the strains were resistant by Autobac but susceptible by the disk test (major discrepancies); and 13.7 percent of the strains were indeterminate by one method and either susceptible or resistant by the other (minor discrepancies). Very major discrepancies were considered serious since the Autobac would be reporting false susceptibility on this particular strain. Therefore, therapy with this antibiotic would be ineffective for this organism. Major discrepancies are not quite as serious since these are false

TABLE 16-II

SELECTION OF A PENICILLIN G CONCENTRATION GIVING AUTOBAC
SUSCEPTIBILITY VALUES CORRELATNG WTH ZONE SIZE
DETERMINATION

Penicillin Concentration (units)	% Interpretive agreement	% Discrepancy*		
		Very major	Major	Minor
0.1	61.0	0.0	20.0	19.0
0.2	71.6	0.0	12.6	15.8
0.4	76.9	1.0	7.4	14.7
0.8	83.2	1.0	2.1	13.7
2.0	82.1	2.1	1.1	14.7
4.0	82.1	3.2	0.0	14.7

*Very major: susceptible by Autobac, resistant by disk
 Major: resistant by Autobac, susceptible by disk
 Minor: Intermediate by one method, susceptible by the other method

resistances reported by Autobac; that is, based on Autobac results this antibiotic would not be selected for therapy, whereas based on the disk test it would be.

A graphical representation which depicts the correlation of the inhibition zone diameters with the light scattering index for each of ninety-five strains using the 0.1 unit penicillin G disk is shown in Figure 16-2. The horizontal lines represent the interpretive breakpoints established by the agar diffusion method. The vertical lines represent the interpretive breakpoints programmed into the Autobac photometer. This figure graphically depicts the results of using too little of the antibiotic. If the disk penicillin content is decreased much below 0.8 unit, about 20 percent of the strains found susceptible by zone size determination appear resistant by the Autobac method. Because too little of the antibiotic was used, the strains were able to grow; thus, the light scattering index (LSI) values are close to zero and the organisms appear resistant. About 19 percent were in the indeterminate range by either of the two methods.

Conversely, the results of using too high a penicillin concentration are shown in Figure 16-3. When using the 4 unit elution disk, one runs the risk of false susceptibility results for penicillin-

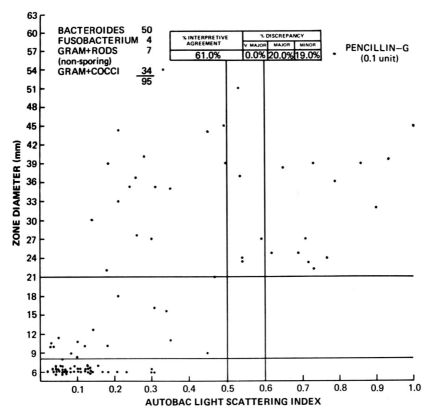

Figure 16-2. Correlation of zone diameters with light scattering index readings for ninety-five anaerobic strains and the 0.1 unit penicillin G disk.

resistant strains. Three of the strains appeared susceptible by Autobac, but resistant by zone size determination. With this high penicillin content, however, the number of major and minor discrepancies decreased.

At the optimal penicillin content (Figure 16-4), only one false susceptible strain appeared. The two methods combined gave a total of 13.7 percent indeterminate results; a majority of these were resistant by Autobac, but indeterminate by zone size measurements. If a large number of strains of a particular species were examined, this wide indeterminate range might be narrowed. In general, these strains have MIC values ranging from indetermi-

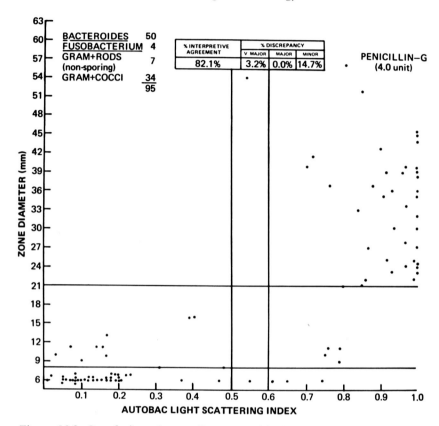

Figure 16-3. Correlation of zone diameters with light scattering index readings for ninety-five anaerobic strains and the 4.0 unit penicillin G disk.

nate to resistant. Two strains were susceptible by zone size measurements, but resistant by Autobac. Based on these data, the authors believe that 0.8 unit penicillin G is the optimal concentration.

The approach taken for penicillin G in these anaerobic studies will now be extended to include other antimicrobial agents.

Developmental work is currently in progress to perfect a system that will make Autobac susceptibility testing of anaerobes a routine procedure for the clinical laboratory. This system will not require an anaerobic chamber. The instrument system will be composed of the present Autobac system with the modifications

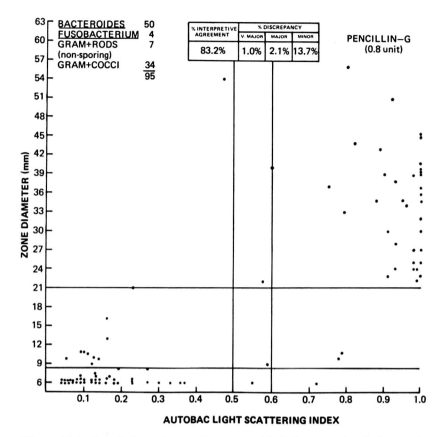

Figure 16-4. Correlation of zone diameters with light scattering index readings for ninety-five anaerobic strains and the 0.8 unit penicillin G disk.

previously mentioned. Two devices have been added. One is a multicompartment airtight chamber called a "cuvette reducing station," used to flush empty cuvettes free of oxygen.

The second is a working station designed to provide the capability for maintaining anaerobic conditions during the manual manipulations required for preparing the standardized inoculum, inoculation of the broth, dispensing the antibiotic disks into the cuvette, and dispensing the inoculated broth into the oxygen-free cuvette. Anaerobic conditions are maintained at this station by the flow of an oxygen-free gas through two steel cannulae. The gas flows out of the cannulae into the broth tube during inoculum standardization and broth inoculation. This station will also supply an oxygen-free gas mixture to the cuvettes during incubation. An additional feature of the station is the electrical sterilization of the cannulae.

The anaerobic Autobac system just discussed will be used in the following manner:

a. Cuvettes are flushed free of oxygen at the cuvette reducing station.

b. The standardized inoculum is prepared at the working station, with the photometer in its anaerobic mode.

c. A panel of up to twelve antibiotic disks is dispensed into the cuvette at the working station.

d. Sterilized prereduced broth is inoculated and distributed into the chamber of the cuvette at the working station.

e. The anaerobic cuvette is incubated for 3 to 6 h.

f. The cuvette is placed in the photometer and the LSI is determined and categorized as resistant, indeterminate, or susceptible by the machine.

REFERENCES

Bauer, A.W., W.M.M. Kirby, J.C. Sherris, and M. Turck: 1966. Antibiotic susceptibility testing by a standardized single disk method. *J Clin Pathol, 45:*493.

Overman, S.B., D.W. Lambe, Jr., and J.V. Bennett: 1974. Proposed standardized method for testing and interpreting susceptibility of *Bacteroides fragilis* to tetracycline. *Antimicrob Ag Chemother, 5:*357.

Sutter, V.L., N.-Y. Kwok, and S.M. Finegold: 1972. Standardized antimicrobial disc susceptibility testing of anaerobic bacteria. I. Susceptibility of *Bacteroides fragilis* to tetracycline. *Appl Microbiol, 23:*268.

Thornsberry, C., T.L. Gavan, J.C. Sherris, A. Balows, J.M. Matsen, L.D. Sabath, F. Schoenknecht, L.D. Thrupp, and J.A. Washington: 1975. Laboratory evaluation of a rapid, automated susceptibility testing system: report of a collaborative study. *Antimicrob Ag Chemother, 7:*466.

Wilkins, T.D., L.V. Holdeman, I.J. Abramson, and W.E.C. Moore: 1972. Standardized single-disc method for antibiotic susceptibility testing of anaerobic bacteria. *Antimicrob Ag Chemother, 1:*451.

Chapter 17

THE USE OF THE DIGITAL COMPUTER IN THE MANAGEMENT AND ANALYSIS OF MICROBIOLOGICAL DATA*

I. J. PFLUG AND R. G. HOLCOMB

Abstract

The digital computer is being used extensively in the Environmental Sterilization Laboratory located in the Space Science Center at the University of Minnesota as a tool for data management, quality assurance, analysis of data, and preparation of tables and graphs. In this report the procedures being used for handling microbiological data are described. Examples are presented illustrating how the computer is being used in a batch mode and in an interactive mode to analyze survivor curves and end-points or quantal response types of microbiological data. Examples of the use of the computer to carry out specific statistical analyses and produce semilogarithmic survivor curves are included. The technical and personnel problems that have to be solved in using an automated system are discussed, together with many benefits that the authors believe have accrued from the use of an automated data handling system in microbiological research in their laboratory.

INTRODUCTION

IN STUDIES of the resistance of microorganisms to heat, chemicals, and radiation stress and in programs to develop biological monitors for sterilization systems, we have been con-

*The developments reported in this chapter were part of the project "Environmental Microbiology as Related to Planetary Quarantine" supported by NASA grant NGL 24-005-160.

The analysis procedures described in this report were developed over a span of several years, with many individuals contributing both to the development and to making the system work. Major contributions in the overall development were made by David Drummond, Ronald Jacobson, Eugene Johnson, and Geraldine Smith.

stantly striving to make both the measurements and the analysis more quantitative. In the data analysis area the authors have benefited greatly from the use of automatic data processing equipment. The computer has been most helpful in bringing coordination and more discipline to the overall laboratory operation.

In this report, the authors will first describe how data are collected and how the computer is used in managing, analyzing, and summarizing the data. Secondly, they will make some general comments on the use of the computer in the analysis and management of microbial destruction data and discuss some personnel and cost considerations in using a computer in the microbiology laboratory.

Experiment Number

An identifying number is assigned to each experiment, and then to the data from that experiment, so that we can easily and positively identify the data from each experiment. To be practical, an experiment numbering system must be simple and easy to use. It must identify the experimenter, specify the day and year on which the experiment is performed, and permit identification of the individual tests of a large experiment or the several experiments carried out on the same day.

The authors use the seven character code illustrated in Table 17-I and have found this rather simple coding system to be very helpful in storing, retrieving, and pooling experimental data. The experiment number is included in both preliminary and final data tables and graphs, so the results are always related to the experiment.

TABLE 17-I

EXPLANATION OF EACH CHARACTER IN THE EXPERIMENT CODE GS5106A

Character No.		Explanation
1 and 2	(GS)	Initials of Project Leader
3	(5)	1975
4 to 6	(106)	April (106th day)
7	(A)	Experiment A of that day

Spore Code

One researcher working with two or three spore cultures has few problems in identifying his cultures and keeping track of experimental results. However, when there are researchers in several laboratories working with dozens of different cultures, it becomes necessary to use a coding system so that identification of all cultures is similar and common communication about them can take place. This system allows all of the spore cultures to be fitted into one identification system. Provision is made for a large number of variations in spore cultures with a continuity of identification. The basic code consists of nine characters, which are designated according to Table 17-II.

TABLE 17-II

EXPLANATION OF EACH CHARACTER IN THE SPORE CODE PBBFL2451

Character No.	Explanation
1 (P)	*Bacillus stearothermophilus* 1518, American Type culture No. 7953
2 (B)	American Type Culture Collection; Rockville, Maryland
3 (B)	Cultured from a water suspension of the lyophilized culture received from ATCC Data cultured: October 1972 Method of culture: Modified Schmidt Cleaning procedure: Insonation, five washings with distilled water
4 (F)	Distilled water at 4°C
5 (L)	Sorensen's Phosphate-Buffer
6 to 8 (245)	Sept. 2 (245th day)
9 (1)	Vial No. 1

Monitoring Temperatures and Calculating Sterilizing Values

In experiments to determine the heat destruction characteristics of microbial spores, precise heat input data are as critical to the experimental results as the microbial survival data. The death of microorganisms is an exponential function of temperature. Small deviations in temperature cause relatively large changes in number of surviving organisms. If 121.1°C is used as the base temperature and the actual temperature is 121.3°C, the microbial death rate will be 4.7 percent greater. If the temperature is 122.1°C, the death rate will be about 26 percent greater. Therefore corrections must be made for the effect of any temperature deviations and for heating or cooling rate. For a heat destruction experiment, thermocouples are placed in each container, miniature retort unit, hot plate unit, or other test apparatus. The output of the thermocouples is monitored by an electronic data monitoring system that provides us with a time-temperature profile of the critical unit in each experiment. This monitoring system (shown diagrammatically in Fig. 17-1) contains timing devices, electronic reference points, and scanning equipment to enable the researcher to keep track simultaneously of up to one hundred separate temperature points. The data are fed to a teletype that

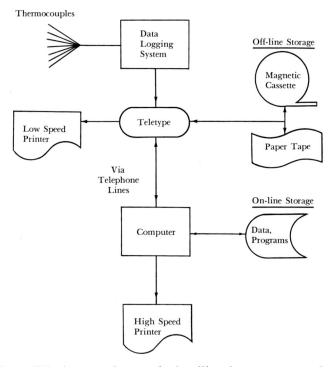

Figure 17-1. Automated system for handling time-temperature data.

can prepare a paper tape or typed copy, or to a digital tape recorder.

The computer is used in several ways to analyze time-temperature data. The data are first sorted and tables prepared of the temperature conditions during the experiment. The sterilizing value of the heat treatment can be calculated as described by Pflug (1973). When heating is carried out in tubes or bottles, heat penetration parameters of the test material in the containers can also be determined. Only automation makes possible the collection and analysis of adequate amounts of these data.

Analysis of Microbial Destruction Data

A major part of the authors' laboratory effort is devoted to measuring the effect of lethal heat test conditions on microbial spore populations, as a function of specific environmental test conditions. Of particular interest is microbial resistance to both dry and wet heat. This report, however, discusses mainly wet heat tests, to determine: (1) the relative sensitivities of different species, (2) the relative sensitivities of different strains and cultures of spores, (3) the effect of the substrate in which the microorganisms are suspended during heating, and (4) the effect of initial concentrations of microbial cells.

All experiments are done in duplicate with each replicate experiment carried out on a different day. Survivor curve tests and FN or quantal tests are conducted. In many cases, the survivor data are further analyzed to determine z-values, with the aim of quantifying the microbial survivor measurement data as far as possible. The use of analytical and statistical measures reduces subjective judgment as much as possible. The objective is to be able to make meaningful judgments regarding the reproducibility of experiments and then the effect of the specific test condition on spore survival.

For the routine analysis of heat destruction data, a semilogarithmic model is used for correlating data thus:

$$log \ N = -U/D + log \ N_o \qquad (1)$$

where:

N = Number of surviving organisms after heating time (U)

N_o = Initial number of organisms
U = Heating time at test temperature
D = Heat resistance parameter

This is probably the most convenient and usable model available today. However, the authors recognize, as discussed by Pflug and Bearman (1972), that all microbial heat destruction data will not fit the semilogarithmic model exactly.

Survivor Curve Analysis

In a survivor curve experiment to evaluate the effect of a heat stress on microbial spores, all conditions except heating time are usually held as constant as possible. The data will be the plate counts of survivors for each heating time, and the unheated control. In the analysis, the survivor data for the unheated controls (N_o) are separated from those for the several heating times. The logarithms of survivor data and heating times are correlated by linear regression analysis (Draper and Smith, 1966) and the slope of the regression line used to estimate the thermal resistance parameter (D). The zero time intercept (Y_o) of the regression line is also calculated and from this the intercept ratio:

$$IR = \log Y_o / \log N_o \qquad (2)$$

The statistical variation of these parameters and the point-to-point D-value of the data are calculated by the computer and printed out in appropriate tables.

The survivor curve data handling system operates in a batch mode with the data on punched cards. A flow-chart of the data handling system is shown in Figure 17-2. The original data for each experimental unit consists of the numbers of colony forming units for each dilution plated (two or three dilutions from each sample, with duplicate plates from each dilution). The best value of the count is determined using laboratory guidelines, and the data are transferred to coding sheets. The microbiologist is now relieved of responsibility for any calculations. The data are then punched on cards, and the computer calculates the desired parameters. The first part of the output, a raw data table as shown in Table 17-III, is verified with the original tabulation sheets. If no

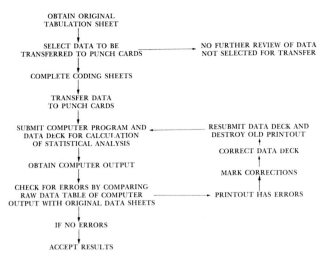

Figure 17-2. Flow chart of survivor curve data handling using card input.

errors are found, all tables in the computer printout are accepted for further study. If errors are detected the data cards are corrected. The corrected data deck is then rerun and a new computer printout obtained. The process is repeated until all known errors are corrected and the final computer printout is accepted for further study.

The second part of the computer printout consists of several tables of summary statistics. The computer can also prepare graphs of results, for example, by drawing a point-to-point line through the mean values, indicating the data point for each experimental unit and the confidence interval for each experimental condition. It will calculate D-values from adjacent experimental conditions, analyses of variance, and intercept ratios. Data sheets and cards are stored for future use.

Quantal Assay Data Analysis

The second type of microbiological data gathered, fraction negative data, illustrates the use of the computer in an interactive mode. With quantal data, instead of recording the number of survivors at each heating time, we record only whether or not

TABLE 17-III
COMPUTER OUT PUT: RAW DATA TABLE

Data table for Experiment number GS5106A.

Experiment started on day-106, year-1975. Sports-PBBT.

Apparatus type-closed, name-sealed glass containers.

Samples conditioned at   0 C,   0 Pct. R.H. for   0 hours.

Samples treated at   121.0 C,   100.00 Pct. R.H. for   times listed.

Comments-4-16-75   Liquid spore storage. Spores deposited 5 min. before heating started. Heated in 5 ml soln. in 18 x 150 ml S.C. T.T. in miniature retort. Spores stored in water for injection for 26 wks. Heated in Sorenson's phosphate buffer ph 7.0.

1↓	2↓	3↓	4↓	5↓	6↓	7↓	8↓	9↓	10↓	11↓	12↓	13↓
1	0	250.00	1.00	51.00	1.00	12750.0000	1	2	2556375.00	6.407625	181	220
2	0	250.00	1.00	51.00	1.00	12750.0000	1	2	2530875.00	6.403271	196	201
3	0	250.00	1.00	51.00	1.00	12750.0000	1	2	2849625.00	6.454788	219	228
4	5.00	250.00	1.00	51.00	1.00	12750.0000	1	2	796875.00	5.901390	62	63
5	5.00	250.00	1.00	51.00	1.00	12750.0000	1	2	828750.00	5.918424	65	65
6	5.00	250.00	1.00	51.00	1.00	12750.0000	1	2	752250.00	5.876362	64	54
7	9.00	500.00	0	0	1.00	500.0000	1	2	77250.00	4.887898	149	160
8	9.00	500.00	0	0	1.00	500.0000	1	2	60500.00	4.781755	125	117
9	9.00	500.00	0	0	1.00	500.0000	1	2	86250.00	4.935759	170	175
10	13.00	125.00	0	0	1.00	125.0000	1	2	5312.50	3.725299	42	43
11	12.00	125.00	0	0	1.00	125.0000	1	2	6312.50	3.800201	45	56
12	13.00	125.00	0	0	1.00	125.0000	1	2	9187.50	3.963197	67	80
13	17.00	50.00	0	0	1.00	50.0000	1	2	250.00	2.397940	6	4
14	17.00	50.00	0	0	1.00	50.0000	1	2	300.00	2.477121	7	5
15	17.00	50.00	0	0	1.00	50.0000	1	2	750.00	2.875061	18	12

Column Titles

(1) = Identification

(2) = Heating period min

(3) = Volume (ml) original suspension (include sample)

(4) = Amount transferred (ml)

(5) = Volume (ml) in receptacle after transfer

(6) = Amount (ml) plated from (5)(from (3) if (5) = 0)

(7) = Dilution factor

(8) = Number of samples per experimental unit

(9) = Number of plates per sample

(10) = Estimated number of survivors in original suspension

(11) = Log_{10} of (10) (disregard if (10) = 0)

(12) = Plate count 1 for sample 1

(13) = Plate count 2 for sample 1

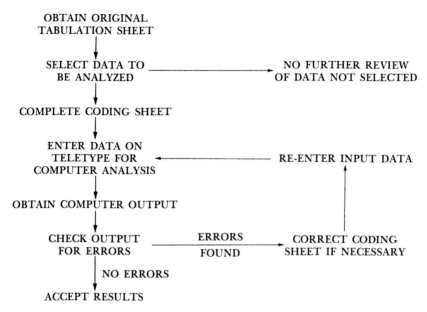

Figure 17.3. Flow chart of fraction negative data handling using teletype input.

there are survivors. At each heating time a positive or negative response for growth after culturing is recorded. The flow of quantal data through the analysis system is shown diagrammatically in Figure 17-3. A sample coding form on which the description of the experiment and the results are recorded prior to analysis is shown in Table 17-IV.

The quantal assay procedure used is based on the Spearman-Karber estimator (Pflug and Holcomb, 1976). The estimate $(\overline{X})$ of heating time is defined by

$$\overline{X} = \sum_{i=1}^{k-1} \left(\frac{U_{i+1} + U_i}{2} \right) \left(\frac{V_{i+1}}{n_{i+1}} - \frac{V_i}{n_i} \right) \qquad (3)$$

where (V_i) is the number of sterile replicate units out of a total of (n_i) units heated for the time (U_i) at test temperature. The estimate $(\overline{X})$ is the expected heating time required to sterilize a randomly selected unit. An estimate of the heat resistance parameter

TABLE 17-IV

QUANTAL ASSAY DATA FORM

Submitted by:

Output for:

Experiment no.:

Insert code:

SPORES

1. Spore code:
2. Storage time:
3. Storage conditions:
4. Carrier system:
5. Time of inoculation:
6. Volume of inoculum:
7. Conditioning:
8. Initial number:

HEATING

11. Heating time units:
12. Time started:
13. Heating system:
14. Heating conditions:
15. Cooling treatment:

RECOVERY

21. Culture media:
22. Tube size:
23. Method:
24. Incubation:

COMMENTS

31. Comments:
32. Comments:
33. Comments:

Data:	*Heating time*	*No. units heated*	*No. sterile*

(D) can be obtained from $(\overline{X})$ and the initial number of organisms per replica (N_o) thus:

$$D = \overline{X} / (.2507 + log\ N_o) \qquad (4)$$

Calculation of z-Values

When comparable D-value data are available at several temperatures, a thermal destruction graph of the logarithm of the D-value versus temperature can be prepared. Using the least squares method, a straight line can be fitted to these data. The negative reciprocal of the slope of this line is an estimate of the z-value, the temperature decrease necessary for the D-value to increase by a factor of 10. The computer can be used in an interactive program exchange in the calculation of z-values from sets of D-values and temperatures and will also calculate the 95 percent confidence intervals.

General Comments on the Use of the Computer

The digital computer has been in wide use in the scientific community for more than fifteen years. There has been experience gathered in many fields concerning both its value and limitations. In spite of limitations resulting from the cost of hardware and man-hours, access to computer facilities in the authors' laboratory has been of very great benefit.

The computer can aid research effectiveness in several areas. One of these is in the storage and manipulation of experimental data. For example, during the past five years, over fifteen hundred survivor curve experiments have been coded, punched, and analyzed. The cards for these experiments are readily accessible for future analysis. If analysis techniques change, new calculations can be made using the stored data without need to repeat the experiments. Should other laboratories desire copies of data in this raw data bank, a duplicate set of cards can be made.

Since an accurate set of information is called for on the coding forms, emphasis is placed on describing and recording all experiments completely. The requirement for completeness in input data to the computer system before analysis can begin acts as an efficient disciplinarian to laboratory personnel. Since the data is

recorded for each experiment in the same way, comparison between sets of data or an examination of long-term results or trends
is facilitated. The storage of experimental data on computer
punch cards has other advantages; for example, the data for individual experiments can be analyzed in different ways. Errors
detected in the raw data printout can easily be corrected.

It sometimes becomes desirable to reduce the physical bulk of
the computer cards of a large data base by transferring the information to, for example, magnetic tapes or discs. Magnetic tape is
the cheaper form of storage, allowing the economical storage of
backup copies of essential data and computer programs. Information can be just as easily communicated to and from computers
with all three systems. Generally, however, information stored on
tapes or discs is less easily transferred between computer sites and
different computer systems than information stored on cards.

The digital computer in an interactive mode is directly useful
to laboratory personnel for carrying out complex or tedious computations and statistical analyses. The authors have prepared a
set of programs that have been tailored to specific data analysis
needs. Fed with data consisting of plate counts, dilution factors,
temperatures, and heating times, the computer will provide estimated survivor and regression curves in seconds.

A great deal of time usually must be invested in writing and
debugging programs before the first data are analyzed. But once
the investment has been made, the technologist can use the program to carry out analyses in an efficient manner, free from human
errors. One factor in this area, however, must be guarded
against. Since an ongoing computer analysis system has an aura of
accuracy and reliability surrounding it, the results are often not
questioned or critically examined. It is important that only data
that are compatible and appropriate for the specific type of analysis
program be entered in the system.

Another area in which the computer can aid in a research effort is in the preparation of suitable graphs and tables for organized laboratory records, and the preparation of reports.

The value of the digital computer described thus far in manip-
ulating data and performing calculations is dependent upon the

size of computer available. The computers available today vary from small laboratory models which are relatively inexpensive and inflexible, to large-scale, very flexible, expensive facilities that service many thousands of users. Several models provide the large-scale storage capacity and programming language availability needed for microbiological data management and analysis work. When purchasing computer time and support services communication can be made with computer card decks (batch mode) and/or over-voice-grade telephone lines using a teletype (either in batch mode or in real-time interactive mode).

Considerations in Setting Up a Computer System

There are many steps involved in deciding whether or not to use computer support in the laboratory (Martin, 1972). The obvious first step is identifying the type and scope of calculations and data handling that need to be performed. By knowing what data management services are required, one can make the necessary tradeoffs in cost, computer flexibility, and support personnel needed. If the calculations are relatively simple and the data sets reasonably small, then a programmable calculator may be all that is required.

For small research groups the purchase of computer hardware itself is generally too costly. The alternative is to purchase computer time on an existing computer facility. Both the grade or quality of service and the variety of services available differ between computer facilities. The service considerations involved include how data may be entered into the computer (cards, magnetic tapes or discs, teletype), what programming languages are available, whether or not there is on-line storage of data and programs, telecommunications cost, freedom of the system from failures, availability of support personnel, output facilities (printing, graph plotters, microfilm printers, card punchers, tape writers) and hours of availability.

For users of batch computers with access through punch cards, some provision must be made for keypunch machine services. If the number of cards to be punched is small and if the need is only occasional it may be most economical to purchase keypunch services. If there is a regular, high volume of data to be punched, it may be preferable to rent or purchase a keypunch machine and hire a keypunch operator. Users of interactive computers will need to lease or purchase a teletype or similar device.

The costs involved in the computer processing of microbiological data are generally difficult to assess. The cost structure is highly dependent on the system used. The costs are dependent, among other things, on the brand of equipment being used, whether the computer facilities are commercially operated or government subsidized, and what types of analyses are being performed.

Some estimates for the data processing costs at the University of Minnesota will be given as a rough guide. The hiring of keypunch services costs about ten cents per card. The computer cards cost less than one-fourth cent apiece. A keypunch machine costs from 18 to 36 thousand dollars depending on the manufacturer. A teletype with a paper tape punch and reader and an accoustical coupler (to communicate over telephone lines with a computer) costs about 15 hundred dollars. A magnetic tape cassette recorder costs about the same amount. Equipment such as keypunches and teletypes can usually be bought or leased for about 5 percent of the purchase price per month. Service contracts for maintenance of such equipment after the warranty period can range from about 5 to 15 percent of the purchase price per annum. Finally computer outputs consisting of the analyses and graph for one survivor curve cost, on the average, a little less than two dollars each.

To obtain the necessary software for data analysis, several options are available. Computer program packages can be purchased which perform a variety of mathematical and statistical procedures. Two well-known packages are the Biomedical Computer Programs (BMD) from the University of California Press, and the IMSL programs from the International Mathematical and Statistical Libraries, Inc., of Houston, Texas. Alternatively, users who require tailor-made programs may hire qualified programmers. Specially written programs oriented to a particular application will result in a greater overall efficiency of computer analysis.

An important factor in the effective use of computers in the laboratory is its integration into the organization. Personnel do not have to understand fully how the computer works or the basis for the analyses, but they do have to accept it as a tool and learn to use it. Effective use is aided by making computer coding forms and outputs as simple as possible. When interactive computer programs are used, it is important that interchanges between man and machine are as close to human conversation as practical. Interactive programs that can solicit information with readily understandable questions help to eliminate confusion and errors.

The cost of maintenance must be added to the initial cost of (1) renting or leasing equipment, (2) providing for computer programs, and (3) training personnel. Mechanical or electrical equipment needs periodic service, and most computer programs also require small changes from time to time.

REFERENCES

Draper, N.R. and H. Smith: 1966. *Applied Regression Analysis.* Wiley, New York.

Martin, J.: 1972. *System Analysis for Data Transmission.* P-H, Englewood Cliffs.

Plug, I.J.: 1973. Heat sterilization. In *Industrial Sterilization.* International Symposium, Amsterdam, 1972, by G.B. Phillips and W.S. Miller. Duke Pr, Durham.

Plfug, I.J. and J.E. Bearman: 1972. Treatment of sterilization process microbial data. In *Environmental Microbiology as Related to Planetary Quarantine,* Progress Report 9, NASA Grant NGL 24-005-160.

Pflug, I.J. and R.G. Holcomb: 1976. Thermal destruction of microorganisms. In *Disinfection, Sterilization, and Preservation,* ed. by S.S. Block. Lea and Febiger, Philadelphia (in press)

Chapter 18

COMPUTER-ASSISTED IDENTIFICATION OF UNKNOWN BACTERIA*

J. G. EDWARDS

Abstract

A method for computer identification of bacteria using test pattern recognition is presented. In addition to present-absent designations for tests, a third character code representing variable results was used for the eighty-one morphological and biochemical tests employed in the program. Unknown and known bacteria test patterns were compared, and various ratios were computed. If the unknown showed an identical or very similar pattern to any one of over 500 known bacteria in the data base, a match was made. The program's logic is illustrated with a flow chart plus examples of output. Results of students' identification of unknown bacteria against the computer's identification are reported.

Introduction

IDENTIFICATION of an isolated bacterium can be a tedious and frustrating task. The eighth edition of *Bergey's Manual of Determinative Bacteriology* (Buchanan and Gibbons, 1974) has helped to standardize the classification of bacteria, but standardization of classification is not *ipso facto* assurance of ease of identification. There are currently two methods used for bacterial identification (Friedman et al., 1973): first, the "flow chart" branching method with two choices per test, which bases its selection of the next test on the results of the previous one until no more branches are left; and second, the "pattern recog-

---

*The author expresses gratitude to Dr. Troy L. Best for his help in the drafting of this manuscript and to Dr. Fred A. Rosenberg and his graduate students at Northeastern University for their cooperation in testing the usefulness of the program.

nition" method which compares a large number of test results for any unknown against known bacteria.

Bergey's Manual of Determinative Bacteriology makes use of the branching method of identification, a method which has the disadvantage of yielding a serious error in identification if only a single test is interpreted incorrectly. Conversely, the pattern recognition method has the advantage of speed and decreased likelihood of gross error even if some tests are misinterpreted. The latter method, however, is rather complex; Friedman, et al. (1973) point out that when fifteen positive-negative tests are considered, more than 2^{15} or 32,768 possible combinations between the unknown and known bacteria would result. Even the most perceptive microbiologist would not be able to visually recognize all of the uncommon patterns.

In order to utilize the accuracy and speed of the pattern recognition method and overcome its complexity, the author adopted the coding system of Friedman et al., applying it to the calculation of various coefficients for the identification of unknown bacteria. The purpose of this chapter then, is to describe a computer program for the identification of bacteria using taxonomic distance and several association coefficients.

Methods

The initial step was to code characters for the Enterobacteriaceae. This group was chosen because the characteristics of its members are well-known and clearly described in the literature (Ewing, 1971). Biochemical tests were chosen from those of Bascomb et al. (1973) and the Difco Laboratories chart (Ewing, 1971) on the basis of ease of performance and frequency of use in teaching or clinical bacteriological laboratories.

The tests used are shown in Table 18-I. The first of these, shape, was given a letter code; the other eighty-one were considered to have three possible outcomes, namely positive, negative, or variable. If a known bacterium was listed as giving a positive result for a test it was coded as 1.0; if listed as negative it was coded as 0.0; and if the frequency of occurrence was < 90 percent, or if the test result was not available, it was coded as 0.5.

TABLE 18-I

BIOCHEMICAL AND MORPHOLOGICAL TESTS USED FOR COMPUTER-ASSISTED
IDENTIFICATION OF UNKNOWN BACTERIA

Except for "shape", use the following codes: Positive = 1; Negative = 0; Variable = 0.5

1. Shape (circle one):

    B = bacillus   V = vibrio

    C = coccus   P = pleomorphic

    Q = coccobacillus   S = spirillum

2. Gram reaction ....

3. Size (5.0 μ long or longer = positive) ....

4. Oxygen requirements ,aerobic = positive) ....

5. Motility (at room T°) ....

6. Colony form (circular = positive) ....

7. Colony elevation (flat or slightly raised = positive) ....

8. Colony margin (entire = positive) ....

9. Blue-green color on eosin methylene blue agar ....

10. Blue-pink color on eosin methylene blue agar ....

11. Colorless colonies on eosin methylene blue agar ....

12. Red colonies on MacConkey agar ....

13. Colorless colonies on MacConkey agar ....

14. Red colonies on salmonella-shigella agar ....

15. Colorless colonies on salmonella-shigella agar ....

16. Red colonies on deoxycholate citrate agar ....

17. Colorless colonies on deoxycholate citrate agar ....

18. Yellow colonies on xylose lysine desoxycholate agar ....

19. Red colonies with black center on xylose lysine deoxycholate agar ...

20. Red colonies on xylose lysine deoxycholate agar ....

41. Capsule present ....

42. Spores formed ....

43. Metachromatic (volutin or polyphosphate) granules present ...

44. Lipid granules present ....

45. Sulfur granules present

46. Dirty colored colonies ....

47. Brown-colored colonies ....

48. Violet-colored colonies ....

49. Green-colored colonies ....

50. Yellow-colored colonies ....

51. Red-colored colonies ....

52. Oxidase present ....

53. Catalase present ....

54. Acid from glucose ....

55. Malonate utilization ....

56. Acid from lactose ....

57. Acid from sucrose ....

58. Acid from mannitol ....

59. Acid from dulcitol ....

60. Acid from xylose ....

61. Acid from salicin ....

62. Acid from inositol ....

63. Acid from sorbitol ....

21. Orange colonies on hektoen-enteric agar ....
22. Pink colonies on hektoen-enteric agar ....
23. Blue-green colonies on hektoen-enteric agar ....
24. Blue-green colonies with black center on hektoen-enteric agar ....
25. Alpha hemolysis on blood agar ....
26. Beta hemolysis on blood agar ....
27. Gamma hemolysis (anhemolysis) on blood agar ....
28. Gelatin liquefaction (at room T°, after 5 days) ....
29. Indole production ....
30. Methyl red test ....
31. Voges-Proskauer test ....
32. Acid production in triple sugar iron agar slant ....
33. Gas production in triple sugar iron agar slant................
34. Acid production in triple sugar iron agar butt ....
35. Gas production in triple sugar iron agar butt ....
36. H₂S production in triple sugar iron agar ....
37. Presence of lysine decarboxylase ....
38. Presence of ornithine decarboxylase ....
39. Presence of phenylalanine deaminase ....
40. Starch hydrolysis ....

64. Acid from arabinose ....
65. Acid from raffinose ....
66. Acid from rhamnose ....
67. Acid from fructose ....
68. Acid from mannose ....
69. Acid from galactose ....
70. Acid from maltose ....
71. Agar digestion ....
72. Cellulose digestion ....
73. Dextrin breakdown ....
74. Acid reaction of litmus milk ....
75. Alkaline reaction of litmus milk ....
76. Reduction of litmus milk ....
77. Coagulation (curd formation) of litmus milk ....
78. Peptonization of litmus milk ....
79. Litmus milk unchanged ....
80. Nitrate reduction ....
81. Urease production ....
82. Citrate utilization ....
83. Give the unknown a code name (up to 35 letters) ....

Prior to the actual computations, these numbers were converted to percentages, i.e. 1.0 = 99 percent, 0.0 = 1 percent, and 0.5 = 50 percent. All eighty-two tests for each of the more than 500 known bacteria were punched onto computer cards. Analyses were performed using the Control Data Corporation Cyber 70 Model 72-14® computer at the Northeastern University Computation Center.

The ability of the program to match an unknown bacterium with one or several known bacteria depends on the calculation of several coefficients based on test results for the unknown and known bacteria listed in the data base. These coefficients are D_{jk}, the taxonomic distance coefficient (Sneath and Sokal, 1973) ; S_1, the simple matching coefficient (Sokal and Michener, 1958) ; S_2, a variation of the Jaccard's coefficient (Sneath and Sokal, 1973) for excluding negative matches; and S_3, a variation of Rogers and Tanimoto's (1960) association coefficient. The purpose of using four different coefficients was to take advantage of as many relationship patterns as possible. Other coefficients may be used, but the simplest are preferable (Bailey, 1967).

The flow chart for the computer program is shown in Figure 18-1 and has been included to guide those readers wishing to write their own programs. Notice that it is presented in three blocks for ease of reading. Starting at the top left the computer reads a number, LASTJ, which is the number of biological tests used. In the present program, LASTJ equals 81. Then the computer reads another number, LASTI, which is equal to the number of known organisms in the data base. Next, the name of each of the eighty-one tests is read in and stored. Continuing, the integer variable, LASTJ, is now set equal to the floating point variable, TOTALN, to permit later calculations. (In Fortran, one should not use integers for division since truncation, not rounding off, results.) Next, each test result for the unknown bacterium is read in, stored, and converted to a percentage. As stated earlier, if a test result is equal to 1.0, it becomes 99.0 percent, if equal to 0.0, 1.0 percent; and if equal to 0.5, 50.0 percent. Next, the test results for the unknown are summed, then each test result is squared and the squares are summed. This brings the reader to

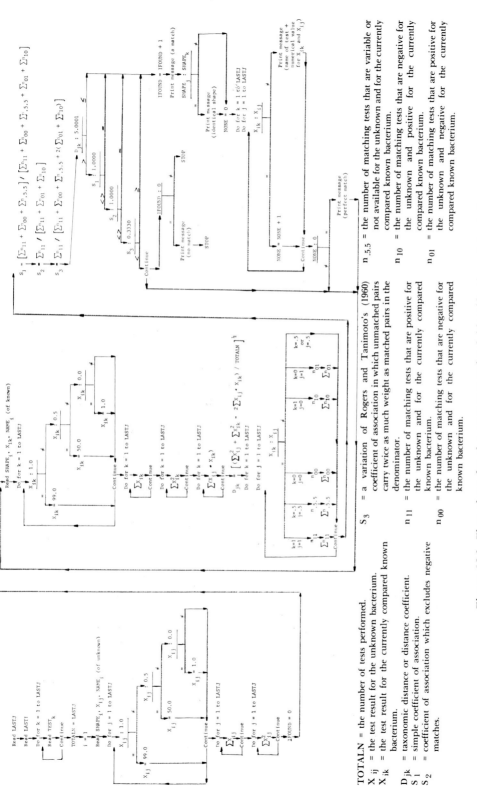

Figure 18-1. Flow chart for the computer-assisted identification of bacteria.

TOTALN = the number of tests performed.
X_{ij} = the test result for the unknown bacterium.
X_{ik} = the test result for the currently compared known bacterium.
D_{jk} = taxonomic distance or distance coefficient.
S_1 = simple coefficient of association.
S_2 = coefficient of association which excludes negative matches.

S_3 = a variation of Rogers and Tanimoto's (1960) coefficient of association in which unmatched pairs carry twice as much weight as matched pairs in the denominator.
n_{11} = the number of matching tests that are positive for the unknown and for the currently compared known bacterium.
n_{00} = the number of matching tests that are negative for the unknown and for the currently compared known bacterium.

$n_{.5.5}$ = the number of matching tests that are variable or not available for the unknown and for the currently compared known bacterium.
n_{10} = the number of matching tests that are negative for the unknown and positive for the currently compared known bacterium.
n_{01} = the number of matching tests that are positive for the unknown and negative for the currently compared known bacterium.

the end of the first block where a counter, called IFOUND, is defined. Now the reader proceeds to the top of the middle block where basically the same procedures that were just outlined for the unknown bacterium are performed for one known bacterium. Three fourths of the way down the middle block, the quantity "D_{jk}", the distance coefficient, is defined. The taxonomic distance between two species of bacteria is equated to the Euclidean distance between two points in an "n"-dimensional space. For a more thorough explanation of this concept, the reader is referred to Sokal and Sneath (1973).

Next comes the heart of the program, that part where each test result for the known bacterium is compared with the corresponding test result for the unknown. The mechanics for this comparison are shown at the bottom of the middle block on the flow chart. It is not difficult to interpret, if the reader follows the legend for abbreviations and keeps in mind that "X_{ij}" designates the test results for the unknown bacterium, while "X_{ik}" represents the corresponding results for the currently compared known bacterium. After sorting out and summing the number of matching and nonmatching positive and negative test results and matching variable test results between the unknown and the known bacterium, one enters the top of the third block segment and calculates three coefficients of association (S_1, S_2, and S_3) as follows.

S_1 is the simple matching coefficient (Sokol and Michener, 1958) in which matching variable test results are considered in the numerator and denominator. This coefficient is the equivalent of the Jaccard coefficient (discussed below) but includes negative matches.

S_2 is the Jaccard coefficient which excludes negative matches (Sneath and Sokal, 1973). Exclusion of negative matches from the computation is considered to be a safe procedure in the field of microbiology by these workers, since a large group of negative test results may be due to an unrecognized metabolic block preventing the expression of many other characters. For example, it may be impossible to decide if enzymes E_n, E_{n+1}, E_{n+2}, etc. are present but not expressed because of lack of regulatory enzyme, E_{n+x}, which is necessary for activity.

S_3 is a variation of Rogers and Tanimoto's (1960) association coefficient in which matching test results are considered only in the denominator, where unmatched pairs carry twice as much weight as matched pairs. The modification consists of not including negative or variable matches in the numerator. This coefficient was chosen because it is recommended for multistate characters (Sneath and Sokal, 1973).

Finally, after a numerical comparison of these coefficients, the computer determines whether the test pattern for the unknown bacterium resembles that for the known bacterium. In regard to the distance coefficient, D_{jk}, the greater the disparity between two organisms, the greater is their taxonomic distance. If any two species are identical in terms of the characters under consideration, their positions will coincide, and the distance between them will be zero. However, because of introduction of the third character state, i.e. "variable", D_{jk} is here compared to 5.0001.

Whether a match is made or not, the computer will return recursively to the top of the second block to search for additional possible matches with the other known organisms in the data base.

Results and Discussion

In order to determine how few of the eighty-one tests needed to be performed before the computer could no longer identify an unknown bacterium, the following experiment was performed: the characteristics for *E. coli* were duplicated and were submitted to the computer, designated as "test *E. coli*". After each computer run, an increasing number of either positive (1.0) or negative (0.0) characters were changed to variable (0.5) to simulate the condition of "not known" or "not performed". The results are shown in Table 18-II.

A number of graduate students assisted in testing the accuracy and usefulness of the program for identifying unknown isolates of bacteria. The students obtained isolates from marine and terrestrial environments in and around Boston, Massachusetts, during October and November, 1975. A comparison was then made between identification based on conventional references (Buchan-

TABLE 18-II

Example to illustrate how few of the 81 tests have to be performed before the computer can no longer match *E. coli* with a "test *E. coli*" for which increasing numbers of positive (1.0) or negative (0.0) test results were changed to variable (0.5):

Number of changes from 1.0 or 0.0 to 0.5	Variable	Positive + Negative	Percentage positive or negative tests	Number of computer matches
0	13	68	85	*E. coli* only
20	33	48	59	*E. coli* only
30	43	38	47	*E. coli* + 1 other
35	48	33	41	*E. coli* + 2 others
40	53	28	45	*E. coli* + 3 others
45	58	23	28	*E. coli* + 9 others

an and Gibbons, 1974; Difco Manual, 1971; Ewing, 1971) and identification based on computer analysis. The results are shown in Table 18-III. Prior to submission of test results for each unknown bacterium to the computer, a mini-program was used to check the correctness of entered data.

Typical computer output is based on one of three conditions being met between the unknown and one or more known bacteria:

(a) A perfect match, e.g.

UNKNOWN BACTERIUM MATCHES CITROBACTER FREUNDII

SHAPES ARE IDENTICAL

ADDITIONAL TESTS THAT WILL DIFFERENTIATE UNKNOWN FROM CITROBACTER FREUNDII:

NONE — ALL TESTS MATCH

(b) The unknown closely resembles a known bacterium in the data base, but one or more tests do not match, e.g.

UNKNOWN MATCHES BACILLUS SUBTILIS

SHAPES ARE IDENTICAL

ADDITIONAL TESTS THAT WILL DIFFERENTIATE UNKNOWN FROM BACILLUS SUBTILIS:

SIZE	50.0	99.0
CAPSULE FORMATION	50.0	1.0
VOLUTIN GRANULES	99.0	50.0
CELLULOSE BREAKDOWN	50.0	1.0

TABLE 18-III

COMPARISON OF STUDENTS' AND COMPUTER'S IDENTIFICATION
OF UNKNOWN BACTERIA

	Student Identification	*Computer's Identification*
1	*Corynebacterium* sp.	*Corynebacterium parvum*
	or	*Corynebacterium kutscheri*
	Arthrobacter sp.	*Mycobacterium tuberculosis*
		Mycobacterium bovis
		Sarcina barkeri
2	*Bacillus* sp.	*Bacillus stearothermophilus*
		Bacillus spaericus
		Sarcina ureae
3	*Bacillus* sp.	*Bacillus badius*
4	?	*Sarcina barkeri*
		Mycobacterium tuberculosis
		Mycobacterium bovis
5	*Staphylococcus aureus* (?)	*Streptococcus pyogenes*
		Streptococcus equisimilis
6	?	*Pseudomonas panicimiliacei*
		P. lignicola
		P. eribotryae
		P. dacunhae
		P. cruciviae
		P. riboflavina
		P. denitrificans
		P. indolixidans
		P. halestorga
		Mima sp.
		Alcaligenes faecalis
		Sphaerotilus natans
		Moraxella sp.
		Acinetobacter lwoffii
		Commamonas percolans
7	*Staphylococcus epidermis*	*Staphylococcus epidermidis*
8	*Proteus vulgaris*	*Proteus vulgaris*
		Beneckea chitinovora
9	?	*Pseudomonas caviae*
		Beneckea chitinovora

(c) The unknown does not match any known bacterium listed, e.g.

UNKNOWN CANNOT BE MATCHED WITH ANY BACTERIUM LISTED

A perfect match is not often obtained. When the unknown closely resembles a known bacterium in the data base, but one or more tests do not match, as under the second condition above.

the nonmatching test results, transformed into percentages for both the unknown and known bacteria, are included in the output. This transformation is done because, in the mathematical calculations that follow, a single negative state (0.0) could reduce all the others to zero.

In regard to the variable character state (0.5) used in this program, the reason for not using the actual numerical frequency of occurrence for a particular character, i.e. 95 percent or 85 percent, etc., is that this frequency varies among laboratories and is not published for all bacteria. According to Bohm (Chapter 2), conventional methods of identifying unknown bacteria result in a 4 percent error within the same laboratory, and an 8 to 15 percent error between laboratories. It is possible that this error is due, in part, to the vagueness of reported frequencies of observance for some characters. Perhaps one of the weakest points of the latest edition of *Bergey's Manual* (Buchanan and Gibbons, 1974) is that it still often describes the occurrence of a test result vaguely. Terms such as: "rarely," "may not be," "generally," "seldom," "little or no," or "to various degrees," are often encountered. This form of test character documentation is perhaps the root of the problem of bacterial identification and leaves the student and many professionals totally exasperated, to say the least.

Since many of the tests used in microbiological characterization do not have a unique result for each species, each test has a percentage of positive-to-negative test results. For example, a 50:50 ratio of positive-to-negative test results provides no information in identifying the bacterium in question. However, as this ratio approaches 100:0 or 0:100, the discriminatory value of the test increases. Thus it is not necessary, nor at times even possible, to know the results of all 82 tests in this program. From the results in Table 18-II, it can be seen that at least 35 percent of the test results (28 out of 81) must be positive or negative for the computer to be successful. As the number of variable test results is increased, i.e. from 53 to 58 out of 81, the pattern for the "test *E. coli*" still matches that for *E. coli*, but the number of other matches increases from 3 to 9. The more tests performed, the fewer the number of possible matches and the greater the likelihood of successful computer identification.

The necessity for using a large number of test results for each bacterium is obvious from the large number of known organisms in the data base. If a small number of tests were studied the discriminatory power of each test would be lowered. Some of the tests provide their greatest usefulness in the identification of a bacterium within a particular group; for example, color of growth on particular types of agar identifies different members of the Enterobacteriaceae (Difco Manual, 1971), while type of hemolysis on blood agar readily distinguishes various pyogenic cocci (Diem and Lentner, 1974).

The results in Table 18-III show that the computer's identification was in good agreement with that of students who identified at least the genus. In four of the nine cases the students were certain of the genus but were uncertain of the species; in other cases, they had no idea on the identity of their unknown. In most instances, the students agreed that the computer's identification plus the listing of unmatching tests were helpful in guiding them to identification. In some cases additional tests were performed; for example, differentiation of *Corynebacterium* sp. from *Mycobacterium* sp. was accomplished using the acid-fast stain.

In only one case (not shown in Table 18-III) was the computer unable to match a student's unknown with any bacterium in the data base. The organism in question did not attack the test sugars prepared in the laboratory, but when inoculated into a commercially prepared strip of some of the same sugars, it fermented almost all of them. The unknown could not be identified using this commercial test strip. Dilution plates showed isolated identical colonies; microscopic examination of a gram's stain from a log growth phase culture showed all gram-negative rods, apparently ruling out contamination or a mixed culture. However, it is possible that contamination occurred during inoculation of the commercially prepared sugar test strip. It is also possible that this discrepancy resulted from thermal degradation of the sugars which in the test were sterilized by autoclaving

(121°C, 15 psi, 15 min.). This perhaps could be overcome by filter-sterilization.

Thus, to encourage progress in facilitating bacterial identification, it will first be necessary to agree on standard methods of media preparation. This in turn would improve test character documentation. Having achieved this groundwork, widespread computer mechanization of identification could become a reality.

The present computer program can be enlarged to include any number of tests for any number of bacteria. It is written in Fortran IV which is compiled by most computers. In addition, a minimum amount of computer storage space is required, enabling the program to be used effectively at institutions having even a small computer.

REFERENCES

Bailey, N.T.: 1967. *The Mathematical Approach to Biology and Medicine.* Wiley, New York, p. 142.

Bascomb, S., S.P. LaPage, M.A. Curtis, and W.R. Wilcox: 1973. Identification of bacteria by computer: Identification of reference strains. *J Gen Microbiol, 77*:291.

Buchanan, R.E. and N.E. Gibbons (ed.): 1974. *Bergey's Manual of Determinative Bacteriology,* 8th ed. Williams and Wilkins, Baltimore.

Diem, K. and C. Lentner (ed.): 1974. *Pathogenic Organisms and Infectious Diseases.* Ciba-Geigy Limited, Basel, Switzerland.

Difco manual, 9th ed. 1971. Difco Labs, Detroit.

Ewing, W.H.: 1971. *Chart for Differentiation of Enterobacteriaceae by Biochemical Tests.* Difco Labs, Detroit.

Friedman, R.B., D. Bruce, J. MacLowry, and V. Brenner. 1973. Computer-assisted identification of bacteria. *Am J Clin Pathol, 60*:395.

Rogers, D.J. and T.T. Tanimoto: 1960. A computer program for classifying plants. *Science, 132*:1115.

Sneath, P.H.A. and R.R. Sokal: 1973. *Numerical Taxonomy,* 2nd ed. WH Freeman, San Francisco.

Sokal, R.R. and C.D. Michener: 1958. A statistical method for evaluating relationships. *U Kansas Sci Bull, 38*:1409.

Chapter 19

APPLICATIONS OF COMPUTERS AND COMPUTER DIAGNOSTIC MODELS IN THE CLINICAL MICROBIOLOGY LABORATORY

J. D. MacLOWRY AND E. A. ROBERTSON

Abstract

Four computerized data bases containing the quantitative antimicrobial susceptibilities, phage lysis patterns, biochemical reactions, and biochemical and demographical epidemiological information on clinical bacterial isolates are described. Using the computer, these data can be retrieved and summarized for easier comprehension. Having the data in computer files permits more elaborate analyses. A stepwise linear discriminant algorithm has been used for bacterial identification and test selection. Maximum likelihood models identify bacteria on the basis of their biochemical reactions or antimicrobial susceptibilities. Relationships between antimicrobials have been studied by correlation, regression, and cluster analysis. Cluster analysis is used to delineate phage lysis groups. The resources required and the costs incurred for computerization are discussed.

INTRODUCTION

THE work presented in this chapter is divided into two broad categories. The first to be discussed will be the data bases which have been developed in the authors' laboratory. The second involves mathematical and diagnostic models as they have been applied in the authors' laboratory.

The data bases were developed for different reasons, at different times, and when different types of computer facilities were available. A number of uses were not anticipated when the data originally were organized. An attempt will be made to distinguish between those things which are being done, as distinct from those things which can be done. Although costs can vary greatly between institutions and depend in large measure on the approach used to solve a particular problem, representative costs will be given.

The authors' laboratory represents the clinical diagnostic microbiology service for a 550-bed research hospital; therefore, the primary responsibility is for patient care work. All of the data bases have been set up to fill a specifically defined clinical need and all of the information is derived from patient care studies. Most of the data represents results which have been generated by routine, standardized laboratory methods.

DISCUSSION
Data Bases
Antimicrobial Susceptibility Data Base

From July 1968 to the present time this laboratory has used the same standardized microtube dilution technique for antimicrobial susceptibility studies (MacLowry et al., 1970). At present there are approximately 37,000 isolates which had minimal inhibitory concentrations (MIC) determined for an average of eleven antibiotics each. The data has been handled in the following way. Each isolate is manually given an internal chronologic laboratory number. This number, along with a coded organism identification number, coded organism site of origin, and the MIC data, are manually entered onto a keypunch form. Batches of these forms are periodically delivered to the central computer facility and punch cards are generated from the forms. The data on these punch cards are transferred to magnetic tape for computer analysis. Two paper copies of the original report are kept, one filed chronologically and the other alphabetically by patient name. Originally, the plan was to record patient identification data on the keypunch form, but this information was too unreliable and subject to too many transcription errors. Therefore, the internal chronologic laboratory number, which is subject to fewer transcription errors, was chosen to connect back to the original patient, albeit in a cumbersome manner. By also keeping a manual alphabetical file on each patient, it is possible to look at other isolates from the same patient.

The usual output from this program is a table. The internal chronologic laboratory number of each isolate is placed at the top of the table. The table is further organized by organism identification (SNOP code number). The body source of the isolates is also noted and the isolates from all body sources are included. Abbreviations of the antibiotics tested are placed along the top of the table and MIC values in the left-hand column. The table gives the number of isolates for each antibiotic at each MIC. The computer will also reproduce the table in percentage form. A detailed tabulation for each organism for the total time interval will be published elsewhere.

Other types of retrievals are possible using this data organization. First, it is possible to select isolates from all body sources, or from a specific body source, or from all sources excluding a specific source. Specified time intervals can be retrieved by utilizing the internal chronologic laboratory number. A final, general type of retrieval can be made using the MIC values. When a specific antibiotic is selected, tables can be generated which include only those isolates at a specific MIC, or only those greater than or less than that MIC. On these retrievals the data is formatted exactly as described above.

The two main types of retrieval which have been used are a print-out for the whole time interval for selected organisms from all body sources and the comparison of various time intervals and body sources one with another. Studies are in progress to look specifically at multiple drug resistance by selecting specific MIC values.

The cost of searching the whole data base is dependent on the size of the data base. To reduce the cost of retrievals, the master file was divided into a set of subfiles each containing a single organism or group of organisms. Printing out the information on 3600 isolates in a subfile costs approximately eleven dollars.

Staphylococcus Aureus Phage Type Data Base

Phage typing of *Staphylococcus aureus* has been performed in this laboratory with a standardized technique under the supervision of one individual since 1957. During this time, there has been little change in the phages used, because they were made in large batches and lypophilized to minimize the necessity of propagation. The file contains approximately 21,000 phage typings, on several thousand different patients between 1967 and 1975. This information was originally kept in notebooks. In 1973 it was decided to enter it into a central computer. The task of putting this information into the computer was extremely arduous and was done by a technologist at a computer input terminal transcribing the information in a specific format. For each isolate, the patient's name, hospital number, date of phage typing, laboratory accession number, body source of the culture, hospital location, and the phage lytic pattern were all entered. Accurate data entry requires diligence and motivation. As the data were entered, computer programs were used to discover input errors. For example, patients' names were alphabetized and the hospital numbers given along with the alphabetic listing. Manual checking revealed misspelling of names and discrepancies in the hospital numbers. The converse was done with identical hospital numbers grouped together and the name displayed for manual review. Some disparities could not be resolved even with review of original data sheets and these entries were stricken from the file.

A representative computer listing of a portion of the file can be seen in Table 19-I. The various fields are indicated along the top of the table. A fixed format file such as this simplifies considerably the task of retrieval and manipulation. Retrieval is usually by hospital number, rather than by patient name. All isolates from a particular hospital location for a specified year can be printed out. Retrievals can be organized by total phage pattern or by any lytic components of a pattern. For example, a listing can be obtained that includes only the pattern 80/81, or it can include all patterns that have 80/81 as part of the total lytic pattern.

The retrievals studied thus far include the phage patterns seen each year, organized either by number of isolates or by individual patients. When organized by patient, the same pattern will only be counted once in a given year regardless of the number of isolates. This is more useful for epidemiologic purposes

TABLE 19-I

PARTIAL PRINTOUT OF *STAPHYLOCOCCUS AUREUS* PHAGE TYPING FILE

Name	Hospital #	Ward	Date	Accn #	Site	Lytic Pattern
J**********U	90-61-32-2	13E	112965	D965	TH	53/86
J**********U	90 61-32-2	13E	083065	K065	TH	6/7/42B/47/52/52A/53/54/75
J**********U	90-61-32-2	13E	083065	K065	MO	6/7/42B/47/52/52A/53/54/75
J**********U	90-61-32-2	13E	092765	A7429	UR	6/7/29/42B/42E/47/52/52A/53/54/75/80
J**********U	90-61-32-2	12E	062165	H119	UR	6/7/29/42B/42E/47/52/52A/53/54/75/80/81
J**********U	90-61-32-2	12E	062165	H121		6/7/29/42B/42E/44A/47/52/52A/53/54/75/80/81
J**********U	90-61-32-2	10E	053165	G174	WD	6/7/29/42B/42E/44A/47/52/52A/53/54/75/80/81

than considering the total number of isolates per year. Another type of retrieval shows the number of isolates from each different phage group each year. Another retrieval places all patients who have had fifteen or more isolates tested into a separate file. This allows study of the changing pattern in a given patient over a long period of time. These studies give some feeling as to how patterns change within a given patient, and also indicate the stability of the phage typing system.

Once the data is entered, verified, and checked, the retrieval costs are relatively modest. Each of the above mentioned retrievals costs approximately thirty dollars. Once the data is in the computer, it is a simple matter to keep the data file current. However, the initial cost of entering sixteen years worth of data is formidable, and one has to be fairly certain that the quality of the data warrants such a tremendous expenditure of energy and money.

Enterobacteriaceae Data Base

In 1972, using information which the authors had accumulated and also using data from standard biochemical reactions (Edwards and Ewing, 1962), it was shown by Friedman et al. (1973) that with the use of a probabilistic mathematical model, excellent diagnosis of Enterobacteriaceae could be performed by computer. In 1973, a much larger data base was given to the laboratory by Mr. P. Janin of Analytab Products, Inc. These data were derived from a 20 test kit, the API 20E®. The 20 test results were converted to a seven digit number, the API Profile Number. A mathematical model was used to evaluate this data.

The maximum likelihood model chosen is a modification of Bayes' theorem. In this model a data matrix is constructed which contains the fraction of tests which have positive results for a specific set of bacteria. The test results for an unknown isolate are then compared to those of each of the known organisms in turn, and a likelihood score is calculated. The known organism which receives the highest score is considered to be the diagnosis of the unknown.

Two additional aspects of this model need to be mentioned.

First, the higher the score obtained, i.e. the closer it is to 1, the greater the possibility that the diagnosis is correct. Conversely, with a lower score, the possibility must be considered that either this is a very rare strain of a particular species, or that the unknown strain was not among the original known organisms considered. This model is used in such a way that a diagnosis must be made from the data given for calculation. It is quite possible that a particular unknown is of a species not included in the data base and a low score reflects this. Second, the closeness of the first and second scores is an indication of the specificity of the diagnosis. If the scores are very close, then the ability to discriminate between the two organisms in question is poor.

An interactive program based on this model for the diagnosis of Enterobacteriaceae has been developed. The data matrix is stored in a central computer. Access is by an interactive terminal in the laboratory. A representative computer session is shown in Table 19-II. This program allows a number of choices as seen at the top of the table. The English language translation of the test pattern can be printed out or suppressed. There are a variety of data bases for the Enterobacteriaceae, and also antibiotic susceptibility data which can be accessed by entering the appropriate command. The month, day, year, coded organism identification, and accession number are then entered. When the API Profile Number is entered, the computer will print a confirmation of the above information and the test pattern. It will calculate the probability and likelihood scores for the possible diagnoses. In addition, any tests against the best diagnoses are listed. Although not seen on this particular output, additional tests can be suggested by the computer which would aid in differentiating the primary diagnosis from each of the secondary diagnoses. Since this program runs on a time sharing computer, the response is virtually instantaneous.

For each interactive session, there is an initial overhead to connect to the computer and load programs. Therefore, a single diagnosis will cost approximately three dollars. However, if a number of diagnoses are performed in rapid succession, the cost per diagnosis falls to approximately forty cents. In practice, this

TABLE 19-II

TERMINAL DIALOGUE FOR INTERACTIVE ENTEROBACTERIACEAE
DIAGNOSIS PROGRAM

WHICH DATA BASE DO YOU WISH TO USE?
1 = NIH, 2 = API
2
ENTER 2 IF YOU WISH TO USE PROFILE NUMBERS
2
ENTER 2 IF YOU WISH TO TEST PATTERN PRINTED OUT
2
ENTERING A CODE NUMBER FOR THE MONTH LETS YOU CHANGE
THE CHOICES YOU HAVE JUST MADE—
    —77 LETS YOU CHANGE DATA BASES
    —88 LETS YOU CHANGE THE AMOUNT OF INFORMATION PRINTED
    —99 ENDS AND UNLOADS THE PROGRAM
MMDDYYSNOPACCN
04077603401111
ENTER API PROFILE #
1404572
ACCN. #: 1111   DATE: 04-07-76

                          * TEST PATTERN *
+ONPG —ADH —LDC —ODC —CIT +H2S —URE —TDA —IND
—VP   —GEL +GLU +MAN —INO +SOR +RHA +SAC +MEL
—AMY +ARA —OXI   NIT

* DIAGNOSIS *	* PROBABILITY *	* LIKELIHOOD *	* LIKELIHOOD RATIO *
CITRO. FREUDNII	0.983	0.15E–01	0.32E + 05
CITRO. FREUDNII 2	0.016	0.25E–03	0.53E + 03
ENT. AGGLOMERANS	0.001	0.11E–04	0.22E + 02

TEST RESULTS AGAINST THIS DIAGNOSIS
  CIT   0.84

---

program is used mainly to identify unusual patterns and is also useful for teaching purposes. It is much cheaper to look up a pattern in a numeric index if this is available, than to have the computer make the diagnosis.

API 20 Profile Number Data Base

Since July 1974, the API 20 Profile Number of each Enterobacteriaceae isolate has been entered through a computer terminal along with the accession number of the specimen and a coded form of the diagnosis made on the organism. At monthly intervals this information is meshed with other administrative information from a separate computer file in which the name of the patient, hospital number, hospital location, source of the specimen, date of the culture, and any appropriate comments are stored. These two files are linked using the common denominator of the specimen accession number. Once this information has been meshed, it is possible to make retrievals of epidemiological interest. A typical printout lists hospital location on the top row. API Profile Numbers are listed in ascending order in the left hand column. Each time a specific profile number is diagnosed from a particular hospital location it is put in the appropriate place in the matrix. The number of isolates are totaled for each profile number and for each location. In the right hand column is listed the percentage of the total for each profile number. Normally these results are generated monthly, semiannually, and yearly. The counts can be organized either by total number of isolates or by total number of different patients for each profile number. Each isolate can be traced directly to a patient using another print-out organized according to hospital location. Within each location, the profile number is listed in ascending order with information on the patient. In addition, a third listing gives the distribution of profile numbers according to the specific bacteriological diagnosis made by the technologist. These data are useful in determining whether or not a specific profile is unusual for a species. Epidemiologically, this information is helpful in assessing unusual isolates in a patient population. The profile numbers are only minimally helpful in terms of tracing nosocomial infections because the specific infections are usually associated with the most common profile numbers. Therefore, it is difficult to know with certainty whether these isolates really are identical. These printouts also present the variation in profile numbers over time. The cost of monthly retrievals is approximately eighteen dollars; that for semiannual retrievals is approximately thirty-five dollars.

Antimicrobial Susceptibility Diagnostic Data Base

In 1972, the MIC data from the antimicrobial susceptibility data base was used for diagnostic purposes (Friedman and Mac-

Lowry, 1973). This was done with a data base of approximately 8,000 organisms including all different species identified in the laboratory. With this relatively crude approach a diagnostic accuracy of 75 percent was attained. The observations which prompted this work were made by technologists in the authors' laboratory, who noticed that certain bacteria could be reliably diagnosed using the antibiogram. Recently, the antimicrobial susceptibility data base was put into a central time sharing computer. Isolates are diagnosed with an interactive program utilizing the probabilistic mathematical model described previously. A considerable amount of work needs to be done to increase its accuracy, particularly by making more diagnostic subgroupings. At the present time, a single diagnosis can be made for approximately two dollars eighty-seven cents; if ten diagnoses are made the cost falls to about eighty-seven cents each. At the present time, the program is not routinely used to diagnose all isolates.

Mathematical and Diagnostic Models

Stepwise Linear Discriminant Analysis

A discriminant function may be thought of as a formula for calculating a weighted average of a set of measurements. Discriminant functions provide a mathematical tool for identifying bacteria on the basis of a set of biochemical measurements. One discriminant function (weighting formula) is established for each diagnostic class by computerized analysis of known isolates (Dixon, 1974). In the case of the Enterobacteriaceae, there would be one discriminant function for *E. coli,* one for *Salmonella typhi,* one for *Proteus vulgaris,* and so on.

To interpret the set of measurements made on a particular bacterial isolate, its test results are evaluated using each of the different discriminant functions in turn. Thus for one isolate, a separate score is obtained for each diagnostic group. The score for *E. coli* measures its similarity to that taxon. Similarly, the score for *P. vulgaris* measures the isolate's similarity to *P. vulgaris.* The unknown isolate is assigned to the diagnostic category receiving the highest score.

An abbreviated set of discriminant functions is shown in Table 19-III. Positive tests are recorded as 1, negative tests are recorded as 2. The upper part of the table shows an abbreviated set of discriminant functions for *E. coli, S. typhi,* and *P. vulgaris.*

TABLE 19-III

DISCRIMINANT FUNCTIONS AND EXAMPLE OF CALCULATIONS
REQUIRED FOR DIAGNOSIS
DISCRIMINANT FUNCTIONS

	—— Test Weighting Factors ——			
Organism	*ONPG*	*Citrate*	*H2S*	*Urea*
E. coli	54	34	109	73
Salmonella typhi	110	15	60	80
Proteus vulgaris	106	31	59	42

EVALUATION OF DISCRIMINANT FUNCTIONS FOR
UNKNOWN ISOLATE

Unknown's pattern	+	—	—	—
	1	2	2	2
E. coli	1 x 54	+ 2 x 34	+ 2 x 109	+ 2 x 73 = 486
Salmonella typhi	1 x 110	+ 2 x 15	+ 2 x 60	+ 2 x 80 = 420
Proteus vulgaris	1 x 106	+ 2 x 31	+ 2 x 59	+ 2 x 42 = 370

The lower portion shows the calculations carried out for an "unknown" isolate (which in this case is actually a typical *E. coli* isolate). To calculate the *E. coli* score, ONPG result is multiplied by 54, the citrate result by 34, the H_2S result by 109, and the urea result by 73. These products are then added together to give the score for *E. coli.* Analogous calculations yield scores for other diagnostic possibilities, the unknown isolate receives a score of 486 for *E. coli,* 420 for *S. typhi,* and 370 for *P. vulgaris.* Thus the isolate would be diagnosed as an *E. coli,* since that identification received the highest score.

A computerized stepwise algorithm may be used to construct the linear discriminant functions. A stepwise linear discriminant algorithm also provides a method for measuring the contribution of individual biochemical tests to the diagnostic accuracy of a battery of tests (Dixon, 1974). Beginning with the actual pattern of test results obtained on a large group of previously identified isolates, tests are chosen sequentially for inclusion in the set of discriminant functions. One discriminant function is constructed for each diagnostic class. At the first step, the test with the greatest potential for discriminating among all the diagnostic groups is chosen. At each succeeding step, the test chosen for addition to the battery of discriminant functions is the one which promises to add the most discriminatory power to the set of tests already selected. The criterion used for test selection makes allowances for correlations between tests. In evaluating a test for inclusion, the algorithm considers only the new information provided by the test. Thus a stepwise linear discriminant algorithm may be used to choose a subset of "most useful" tests from a larger set of tests.

To evaluate the usefulness of a stepwise linear discriminant algorithm for bacterial identification, the authors have used the method to interpret the results of twenty-one biochemical tests performed on approximately 27,000 Enterobacteriaceae which were studied with the API 20E kit. All but 4.9 percent of the isolates were assigned to the correct genus. By way of comparison, the Bayesian maximum likelihood model discussed earlier had a somewhat lower error rate of 3.2 percent when applied to the same isolates. In practice, more complete information is available from reference and experimental sources, and this information is routinely incorporated into the probability matrix used in the maximum likelihood model (Robertson and MacLowry, 1974). When the supplemented probability matrix was used, the maximum likelihood model had an error rate of only 0.6 percent.

To evaluate the stepwise algorithm as a method of test selection, 27,000 isolates were rediagnosed using the following subsets of the twenty-one tests:

a. the first ten tests considered most useful by the stepwise algorithm

b. the last eleven tests considered least useful by the stepwise algorithm

c. the ten tests included in the commercially available API 10S kit (an abbreviated version of the API 20E) .

Table 19-IV shows the various combinations of tests which were used. Since the maximum likelihood model achieves greater

TABLE 19-IV

BIOCHEMICAL TESTS ANALYZED BY THE STEPWISE LINEAR
DISCRIMINANT ALGORITHM

	Biochemical Tests
* ‡	ONPG
*	Arginine dihydrolase
† ‡	Lysine decarboxylase
† ‡	Ornithine decarboxylase
* ‡	Citrate
* ‡	Hydrogen Sulfide
* ‡	Urease
* ‡	Tryptophan deaminase
* ‡	Indole
†	Voges-Proskauer
*	Gelatin
† ‡	Glucose
*	Mannitol
*	Inositol
†	Sorbitol
†	Rhamnose
†	Sucrose
†	Melibiose
‡	Amygdalin
† ‡	Arabinose
†	Oxidase

* First ten tests selected by stepwise algorithm as most useful

† Last eleven tests selected by stepwise algorithm as least useful

‡ Ten tests included in API/10S

diagnostic accuracy, it was used to evaluate the effectiveness of the stepwise test selection procedures. Using all twenty-one tests, 3.2 percent of the isolates were misidentified at the genus level. Using only the first ten tests selected by the stepwise algorithm, the error rate was 7.9 percent. A comparable error rate of 8.4 percent was obtained with the ten tests of the API 10S kit. The error rate of the last eleven tests chosen by the stepwise algorthim rose to 34.4 percent, confirming that the algorithm had indeed included the more useful tests and excluded the less useful ones. The problem of test selection is not a trivial one: over 300,000 different sets of ten tests can be chosen from an original set of twenty-one tests.

The stepwise linear discriminant algorithm offers several advantages:

a. It can automatically interpret a large number of tests simultaneously.
b. It measures the relative diagnostic importance of each test.
c. It automatically weights each test according to its relative diagnostic importance.
d. It automatically allows for linear correlations between tests, thus making allowance for duplicate information provided by different tests.
c. It automatically weighs each test according to its relative variables.

Several theoretical considerations potentially limit the use of linear discriminant functions.

a. Construction of the discriminant functions requires a complete set of test results for each reference organism. A tabulation of the percent positive results for each test for each species is not sufficient.
b. The discriminant functions are constructed for use with a specific set of tests. No test in the set may be omitted. The results of additional tests not included in the set are ignored.
c. The algorithm used to construct the discriminant functions assumes that the results of each of the tests are continuous normally distributed variables.
d. Furthermore, it is assumed that a given test has the same correlations with other tests in each of the diagnostic groups.

Clearly, the assumptions of continuity, normality, and equal standard deviations in every group are grossly violated by tests scored as only positive or negative. Nevertheless, the model still provides useful information, even though its underlying assumptions are not fully met.

Estimation of Antimicrobial Susceptibility
by Means of Regression Models

When several variables are measured on the same isolate, it is often possible to estimate the value of one by using the measured values of the others. Regression analysis provides techniques for doing this. A model is constructed which expresses the variable to be predicted as a function of the measured variables. The coefficients of each term included in the model are chosen to minimize the discrepancies between the predicted value and the observed values. Specifically, the coefficients are chosen to minimize the sum of the squares of these discrepancies.

It is often not economically or technically feasible to measure all variables of interest on each isolate. Thus, while there are dozens of antimicrobial agents available, a clinical isolate is usually tested with only eight to twelve drugs. It would be useful to be able to predict the susceptibility of the isolate to those agents which are not tested. Also, if the results of testing one of the drugs included in the routine screening panel could be accurately predicted, given the results of the other drugs tested, it might be desirable to drop the redundant drug and substitute for it a drug whose susceptibility could not be adequately predicted without testing. Regression analysis might also provide a check on the internal consistency of a set of measurements, and thus provide a quality control check for detecting errors in test results. Table 19-V illustrates one regression model and the

TABLE 19-V

ESTIMATION OF PENICILLIN SUSCEPTIBILITY BY LINEAR REGRESSION

Model:
$$PEN = 2.82 + 0.43\ AMP + 0.19\ CEPH + 0.08\ SULF$$
E. coli Isolate #48:

Ampicillin (AMP)	=	18
Cephalothin (CEPH)	=	18
Sulfa (SULF)	=	13

Prediction:
$$PEN = 2.82 + 0.43 \times 18 + 0.19 \times 18 + 0.08 \times 3$$
$$= 2.82 + 7.74 \quad + 3.42 \quad + 0.24$$
$$= 14.22$$
Measured Penicillin (PEN) = 14

calculations involved in predicting the penicillin susceptibility of an *E. coli* isolate. "Antimicrobial level numbers" corresponding roughly to the tube numbers of a series of twelve dilutions were used in the models. A logarithmic function of the MIC values was, therefore, used. The penicillin level is estimated from the equation:

penicillin = 2.82 + 0.43 ampicillin + 0.19 cephalothin + 0.08 sulfa.

In Table 19-V, this leads to a predicted penicillin level of 14.22. For this isolate the measured penicillin level was 14. When 636 *E. coli* isolates were studied, this linear regression model was able to explain 65.6 percent of the variance of the observed penicillin levels.

More complicated nonlinear models, which included higher order terms such as the squares of drug levels and the product of the levels of a pair of drugs were also examined. In a typical model also including three second-order terms, for example, minus .05 ampicillin^2 plus .04 ampicillin times cephalothin minus .05 cephalothin^2 the model succeeded in accounting for 83 percent of the variance in penicillin levels. Table 19-VI compares the performance of these two models in more detail.

TABLE 19-VI

ACCURACY OF PREDICTION OF PENICILLIN SUSCEPTIBILITIES OF 636 *E. COLI* ISOLATES USING A LINEAR AND A SECOND ORDER REGRESSION MODEL

Terms in Regression Model	Kind of Model	% Variance Explained	% Within Levels		
			±1	±2	±3
AMP, CEPH, SULF	Linear	66	73	93	99.4
AMP, CEPH, SULF, AMPxAMP AMPxCEPH, CEPHxCEPH	Second order	83	89	96	99.7

The success of regression models in predicting antimicrobial susceptibilities depends on the drugs and organisms involved. Table 19-VII summarizes the performance of linear regression models in predicting the susceptibilities of 263 *Staphylococcus aureus* to twelve different antimicrobials. For each drug a model

TABLE 19-VII

PREDICTION OF ANTIMICROBIAL SUSCEPTIBILITIES* OF *S. AUREUS* ISOLATES USING LINEAR REGRESSION MODELS

Antimicrobial	% of Variance Explained by Regression on Other 11
Ampicillin	92
Penicillin	91
Oxacillin	88
Cephalothin	87
Carbenicillin	67
Erythromycin	55
Kanamycin	55
Gentamicin	50
Tetracycline	42
Lincomycin	39
Chloramphenicol	31
Nitrofurantion	19

$p < .0001$ for all regressions
*Only linear terms included.

was constructed using the other eleven drugs. For ampicillin, penicillin, oxacillin, and cephalothin, linear regression models explained approximately 90 percent of the variance in observed levels. Smaller fractions of the variance were explained for other drugs. Even for nitrofurantoin, for which only 19 percent of the variance was explained, the relation between the predicted and measured susceptibilities was statistically significant at the .0001 level. The susceptibility of an individual isolate to nitrofurantoin could not, however, be predicted with this particular regression model. The standard error of the estimated nitrofurantoin level was 1.82 levels. Nevertheless, the small but definite relationship in nitrofurantoin resistance with resistance to the other antimicrobials tested may be of interest from the standpoint of hospital epidemiology or bacterial population genetics.

Correlation Analysis of Antimicrobial Susceptibility Patterns

The correlation coefficient provides a simple method for detecting a relationship between a pair of variables and for measuring the magnitude of that relationship. A high correlation between the susceptibilities to two different drugs may suggest that they are chemically similar, that they have similar mechanisms of action, or that resistance to both of them is carried on the same R factor. In addition, correlation coefficients may be used to guide the choice of variables to be included in regression models. They have also been used to check on the independence of variables used in the maximum likelihood diagnostic model described earlier. In general, correlation coefficients provide a convenient method for screening a large body of data for relationships between variables, and are readily calculated using available computer programs.

When 777 isolates of *S. aureus* were analyzed, the correlation between penicillin and ampicillin was .94. This presumably reflects the similarity of structure, mechanism of action, and mode of inactivation of the two drugs. Similarly, for 469 isolates of *Pseudomonas aeruginosa* the coefficient of correlation for the aminoglycoside antibiotics gentamicin and kanamycin was .56. This correlation is significant with a p value of .0001. For isolates of *E. coli,* highly significant correlations were obtained between tetracycline and streptomycin, sulfa, ampicillin, chloramphenicol, and kanamycin. Perhaps this suggests that genes for resistance to these drugs travel on the same R factors. The correlation of tetracyline with colistin, nitrofurantoin, and nalidixic acid were not statistically significant.

The correlation between a given pair of antibiotics is different for different species. For example, the correlation between tetracycline and chloramphenicol was .54 for *P. aeruginosa,* .37 for *E. coli,* and only .20 for *S. aureus.* These correlation differences among taxa reflect different cell-wall permeabilities, R factors, mechanisms of antibiotic inactivation, etc.

When the population used to calculate the correlation coefficients is heterogeneous, correlations may be found which are not present when only one homogeneous subgroup is analyzed. Table 19-VIII illustrates this for the correlation between penicillin and nitrofurantoin. These data indicate that if the identity of an isolate were known, knowledge of the penicillin level would

TABLE 19-VIII

CORRELATIONS BETWEEN PENICILLIN AND NITROFURANTOIN
SUSCEPTIBILITIES WITHIN INDIVIDUAL SPECIES AND FOR A
GROUP ENCOMPASSING ALL ISOLATES TESTED

Organism	N	Correlation	P
E. coli	822	.04	.24
P. aeruginosa	494	.04	0.66
S. aureus	545	.10	.01
All isolates	4029	.46	.0001

N = number tested.   P = p value.

not improve the estimate of the nitrofurantoin level. If, however, the identity of the isolate were unknown, knowing the level of penicillin susceptibility would significantly reduce the uncertainty involved in estimating the nitrofurantoin susceptibility.

It is interesting that virtually all of the correlations which the authors have observed have been positive, indicating that if an isolate is resistant to one drug, it has an increased chance of being resistant to others.

The Use of Cluster Analysis to Demonstrate Relationships Among Groups of Variables

By examining the interrelations of a group of variables, clusters of biologically related variables may be identified. These clusters might represent, for example, antibiotics with a common mechanism of action, drugs for which the resistance genes are carried on the same R factor, or *Staphylococcus* bacteriophages belonging to the same lytic group.

A clustering method (Dixon, 1975b) the authors have used calculates the correlation coefficients of each variable with every other variable. The clustering process starts by finding the pair of variables with the highest correlation. These are joined to form the first cluster. At each succeeding step a new cluster is formed by joining either a pair of variables, a pair of previously formed clusters, or a variable and a cluster. Eventually every variable is joined directly or indirectly to every other variable. The results of the clustering process are presented in the form of a tree diagram. A tree diagram shows which variables are included in each cluster, and how the clusters are joined to each other.

A tree diagram which summarizes the clustering process for eight antimicrobial drugs tested against a series of *P. aeruginosa* isolates is shown in Figure 19-1. The first cluster was formed by joining penicillin and ampicillin, the two variables with the highest correlation. Next, cephalothin was added to the penicillin-ampicillin cluster. Then a new cluster was formed by joining tetracycline and chloramphenicol. Next, another new cluster was formed by joining streptomycin and kanamycin. Then the tetracycline-chloramphenicol cluster was merged with the streptomycin-kanamycin cluster to form one large cluster. Next, the cluster formed by the first three drugs was joined to the cluster formed by the last four drugs. Finally, colistin was joined to the large cluster containing the other seven drugs.

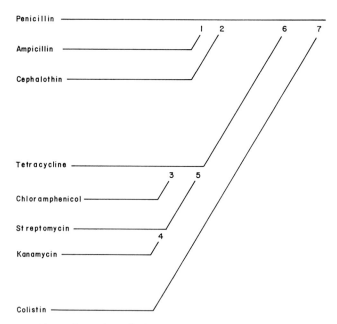

Figure 19-1. Clustering of antimicrobials on the basis of susceptibility results. The numbers at the branch points indicate the order in which the clusters were formed.

Notice that the three antibiotics which inhibit cell wall synthesis (penicillin, ampicillin, and cephalothin) fell into one cluster. Tetracycline and chloramphenicol, which both inhibit protein synthesis, clustered together. The two aminoglycosides, streptomycin and kanamycin, also clustered together. The aminoglycosides were joined to the cluster containing tetracycline and chloramphenicol. Colistin, which has a mechanism of action different from any of the other antimicrobials tested, was excluded from all clusters until the final step.

The same kind of analysis has been applied to *S. aureus* phage lysis patterns. Each typing phage is treated as a separate variable. Each staphylococcal isolate tested is considered a separate case. The complex tree diagram shown in Figure 19-2 resulted from clustering nineteen phages used to type 834 isolates of *S. aureus*

TABLE 19-IX

DELINEATION OF *S. AUREUS* PHAGE LYTIC GROUPS BY CLUSTER
ANALYSIS OF VARIABLES

834 isolates in 1958; 450 in 1973	
Lytic group I:	29/52/52A(79)/80
Cluster group 1:	29/52/52A /80/81*
Lytic group II:	(34)/3B/3C/55/(71)
Cluster group 2:	3B/3C/55
Lytic group III:	6/7/42E/47/53/54/75/77/83A
Cluster group 3:	6/7/42E/47/53/54/75/77/83

*81 not used in 1973 analysis.
Phages 3A and 79 not used at NIH; phage 71 not included in this analysis.

during the year 1958. Although the tree is quite complex and many subclusters are evident, three large clusters corresponding to lytic groups I, II, and III may be discerned. Table 19-IX compares classical lytic groups with the clusters found in the 1958 and 1973 isolates. Lytic group I and cluster group 1 show fairly close agreement. Phages 3A and 79 are not used in the authors' laboratory. For the phages used and included in this analysis, lytic group II and cluster group 2 correspond exactly. Cluster group 3 shows excellent corespondence with lytic group III.

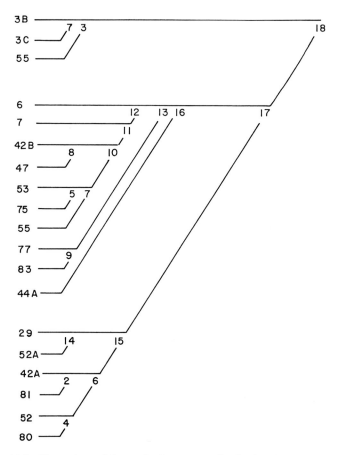

Figure 19-2. Clustering of bacteriophages on the basis of lytic response of 834 *S. aureus* isolates tested during 1958. The numbers at the branch points indicate the order in which the clusters were formed.

By examining the within-cluster similarity of the clusters formed, it is possible to see how tight and clearcut the clusters are. Using nineteen phages as variables, a total of eighteen clusters are eventually formed. Table 19-X compares the within-cluster similarities (on a scale of 0 to 100) of the eighteen clusters formed in 1958 with the similarities of clusters formed in 1973. In 1958, the first ten clusters formed were tight and relatively clear-cut; they all had within-group similarities of 80 or greater. In 1973

TABLE 19-X

SIMILIARITIES OF *S. AUREUS* PHAGE LYTIC GROUPS BY CLUSTER
ANALYSIS OF VARIABLES

40 - 60	Number of Clusters	
Similarity When	1958	1973
Cluster Formed	10	1
80 - 100	4	16
60 - 80	4	1

the groups were less clear-cut; only one group had an average similarity greater than 80. Most of the 1973 clusters had intermediate similarity values in the 60 to 80 range.

Resources Available and Costs Incurred
for Microbiological Computerization

The computerized analysis of microbiological data requires a substantial initial investment for data preparation, entry, editing, correlation, and program development and procurement. Once the data is in a fully edited, computer-readable file, the additional cost of elaborate analysis is modest.

Data preparation, entry, editing and correction, and preliminary analyses are the most time-consuming activities; for example, the process of computerizing the phage-typing data took approximately two years. The initial data entry required several man-months of clerical time, while editing and correcting occupied several calendar months. Ambiguous patient indentifications had to be resolved, missing patient numbers supplied, and extraneous and hopelessly garbled records removed.

When constructing a computer file from handwritten records of patient data, identifying individual patients in an accurate and consistent manner requires much effort. In the case of the phage data, the handwritten records span almost two decades. Deciding which records belong to the same person is difficult. Consider records on the following hypothetical patient names.

J. Jones
J. E. Jones
Jerry Jones

Jerry E. Jones
Jerry Janes
Jerre Jones
Gerry E. Jones

Which of these records should be considered to belong to the same patient? When recorded, the hospital number is of considerable help, but it is subject to frequent transcription errors. Over a period of time certain patients receive two or more different hospital numbers. Also, names change due to marriage. Hospital location, date of birth, sex, and similar clues can help if available.

The choice was made to base patient identification on the hospital's numbering system. Fortunately, this laboratory has available a computer file of all the hospital numbers which have been assigned. This file includes the correct name and simple demographic information on the patient who has been assigned each number. Since 1972, the hospital has incorporated a check digit into each hospital number. A computer evaluation of this check digit makes it possible to detect most transcription errors. In addition, a computerized on-line Soundex file is available which allows researchers to obtain a list of the names and numbers of patients whose last names are similar to a particular name. For example, starting with Jones, the names and numbers for all the Jones, Janes, Janns, etc. can be obtained. Using these tools, it was possible to resolve most of the patient identification problems, but the process required several man-months of effort by professionals and technologists.

W. J. Dixon, who edits the BMD series of biostatistical computer programs, estimates that preliminary steps "can be expected to demand at least ten times more effort than an analysis of the main research hypothesis for which the study was launched" (Dixon, 1975b). Clearly he speaks from experience.

The following illustrates typical on-going costs of computer use. Terminal rental costs 130 dollars per month for each terminal. Program and data storage for the interactive diagnosis program costs less than ten dollars per month. Terminal use costs three dollars per hour. Actual computer time used costs two dollars per second.

Table 19-XI shows the execution costs of the diagnostic programs on a per isolate basis. A large batch run of a discriminant function program, in which over 27,000 isolates were diagnosed, costs less than two cents per isolate. A batch run of the maximum

TABLE 19-XI

COMPUTER CHARGES PER ISOLATE DIAGNOSED FOR DIFFERENT PROGRAMS

Discriminant Function	(Batch of 27,847 isolates)	1.43¢/isolate
Maximum Likelihood	(Batch of 1,200 isolates)	8¢/isolate
Maximum Likelihood	(Interactive Terminal)	
1 isolate		$2.98/isolate
10 isolates		$ .41/isolate
Incremental Cost		$ .12/isolate

likelihood program costs eight cents per isolate. The interactive version of the maximum likelihood program is more costly due to the overhead of the time-sharing operating system and the charges for terminal time. Table 19-XII illustrates that the cost of signing on to the system and loading the program is substantial. After that, the incremental cost is modest.

Program development costs are substantial, typically running in

TABLE XII

ANALYSIS OF COMPUTER CHARGES FOR INTERACTIVE ENTEROBACTERIACEAE DIAGNOSIS PROGRAM

Operations	CPU‡ Time	Elapsed Time	Total Charge
Logon*	.16 sec	.25 min	$0.87
Logon, load program†	1.02 sec	0.8 min	$2.79
Logon, load program, diagnose 1 isolate	1.04 sec	2.5 min	$2.98
Logon, load program, diagnose 10 isolates	1.40 sec	11 min	$4.14

*Refers to signing on to the computer from the terminal.

†Refers to retrieving the program from disk storage and placing it into the main computer memory.

‡Refers to the actual time used by the central processing unit of the computer.

the neighborhood of five dollars per line of the completed program. Until recently all programs were written at the laboratory; now, whenever feasible, prepackaged statistical programs are used. Two statistical packages that have been used are the BMD Biochemical Computer Program (Dixon, 1974; 1975b) developed at the University of California at Los Angeles and the SAS Statistical Analysis System (Jolayne Service, 1972) developed at North Carolina State University. These packages have programs for most of the standard statistical procedures and many of the exotic ones. The authors and coworkers have used these programs for correlation, regression, cluster, and discriminant analysis. In addition, these packages have programs for summarizing, plotting, and listing data. The BMD programs are written in the Fortran language. They can be installed on computers which have at least 128,000 bytes of core memory. The SAS programs are written to run on IBM computers under the OS operating system. As a minimum, they require an IBM 360 computer with at least 140,000 bytes of core storage available to the user.

REFERENCES

Dixon, W.J. (ed.): 1974. *BMD Biomedical Computer Programs.* p. 233. University of California Press, Los Angeles.

Dixon, W.J.: 1975a. First steps. *BMD Communications,* No. 4, p. 1.

Dixon, W.J. (ed.): 1975b. *BMDP Biomedical Computer Programs.* Univsity of California Pres, Los Angeles, pb 307-321.

Edwards, P.R. and W.H. Ewing: 1962. *Identification of Enterobacteriaceae,* 2nd ed. Burgess Publishing Company, Minneapolis.

Friedman, R.B., D. Bruce, J.D. MacLowry, and V. Brenner: 1973. Computer-assisted identification of bacteria. *Am J Clin Path, 60:*395.

Friedman, R.B. and J.D. MacLowry: 1973. Computer identification of bacteria on the basis of their antibiotic susceptibility patterns. *Appl Microbiol, 26:*314.

Jolayne Service: 1972. *A User's Guide to the Statistical Analysis,* pp. 94 North Carolina State University, Raleigh.

MacLowry, J.D., M.J. Jaqua, and S.T. Selepak: 1970. Detailed methodology and implementation of a semiautomated serial dilution microtechnique for antimicrobial susceptibility testing. *Appl Microbiol, 20:*46.

Robertson, E.A. and J.D. MacLowry: 1974. Mathematical analysis of the API Enteric 20 Profile Register using a computer diagnostic model. *Appl Microbiol, 4:*691.

INDEX

319

Date Due